THE BODY LANGUAGE AND EMOTION OF DOGS

THE BODY LANGUAGE AND EMOTION OF DOGS

A Practical Guide to the Physical and Behavioral Displays Owners and Dogs Exchange and How to Use Them to Create a Lasting Bond

Myrna M. Milani,
D.V.M.

WILLIAM MORROW AND COMPANY, INC.
New York

Library of Congress Cataloging-in-Publication Data

Milani, Myrna M.
   The body language and emotion of dogs.

   1. Dogs—Behavior.   2. Dog owners—Psychology.
3. Nonverbal communication (Psychology)   4. Emotions.
I. Title.
SF433.M55   1986   636.7   86-8449
ISBN 0-688-06239-3

Printed in the United States of America

      3   4   5   6   7   8   9   10

BOOK DESIGN BY BERNARD SCHLEIFER

to Mom and Dad
with love
and thanks for the first dog
that changed my life

ACKNOWLEDGMENTS

The following contributed to this book on many different levels and made it a particularly satisfying and cherished endeavor:

Michael Snell, whose support and helpful criticism throughout the evolution and realization of this book was invaluable

Job Michael Evans, whose unique and extensive insights into training expanded my own views tremendously

Brian Smith, whose special bond with our dogs, so different from my own yet no less strong, first made me aware of the infinite forms love and caring can take

Nylon, Dacron, Spike, Dufie, Troubles, Tiffany, Cider, Chewie, Kelly, Sage, Max, Billy William III, and all the other incredible canines who share and shared so much

CONTENTS

INTRODUCTION

DURING THE FIRST TEN YEARS I devoted my energies to the practice of small-animal medicine, I acquired tremendous respect for the critical role body language and emotion play in the relationship between owner and dog. The most thorough diagnostic work-up using the most sophisticated equipment and leading to the most effective cure has no meaning if the owner is afraid to medicate his or her own dog. All that technology becomes an albatross around the neck of the terminally ill pet whose owners refuse to let it die because "We love him too much." And surely every practitioner has been pressured by the guilt-ridden owner to instantly repair a puppy's leg, broken in a fit of rage and impatience when the "spiteful" pup was flung against the wall because it soiled the rug again.

Initially I began my observations strictly for my own enlightenment. I wanted to know why sure-fire medical and/or training regimes didn't work for certain owners and dogs. Why did Ms. Corcoran still allow Hornsby to intimidate her even though they'd been through obedience classes twice? Didn't the Aubuchons realize there were other ways to express love for Misty besides feeding her those rich treats that were wreaking havoc with her digestive tract? How can we ever hope to break this wretched cycle of destruction, punishment, and guilt if

John Hascome insists it's mean to den Pumpkin, "regardless how hatefully she behaves" when left alone?

As I endeavored to answer these questions for myself, I realized that some of the knowledge about dogs that I took for granted wasn't common knowledge at all. Basic behavioral concepts, such as the social nature of dogs, dominance, submission, isolation behavior, territoriality, fear, and aggression, that govern every move our dogs make were unknown to even the most concerned owners. When I had time to share this information, owners were invariably pleased and appreciative: "You mean he's not messing in my bedroom because he *hates* me?" shouted one client in obvious delight after I'd explained the nuances of territorial marking. "My God, that's wonderful! I'm so relieved."

Unfortunately, many times immediate medical or behavioral problems require the veterinarian or trainer to do something right away, leaving little time for detailed discussions of the kind that appear in the following pages. However, the objective, unemotional meaning of body-language displays, the emotional motivation we assign them, and our subsequent response is what dog ownership is all about. If we believe our dogs spiteful, impatient, or unhappy, and irreversibly so, we'll have spiteful, impatient, unhappy dogs—and everything we do together with our dogs will reflect those beliefs.

A client once asked me why another client's pets were always so cheerful, patient, and well-behaved whereas her own, even if from the same litter, were invariably unhappy, spiteful, and ill-behaved. Rather than give that answer now, let's work it out together.

While all owners have lists of emotions that affect their unique relationships with their pets, I've chosen to discuss those that create the most common, and often the worst, problems for both owner and pet. However, this is more than a guide to problem solving. It's an exploration of the dog's exquisitely evolved natural behavioral displays and how they complement and undermine its relationships with humankind. And it's a study of the powerful role human emotions play as we seek to bond with the only species that trusts us and wants to be with us enough to get so close.

THE BODY
LANGUAGE
AND EMOTION
OF DOGS

1

SEEING IS NOT BELIEVING

Bob Kalish had just whistled for his sheepdog, Shaggy, to terminate their romp in the park when a group of picnickers asked for directions. As Bob stood talking with the group, a gray and white blur shot out of the dense undergrowth and barreled down on them like a cannonball.

"Excuse me," said Bob, stepping apart from the group. As the others watched in horror and then amazement, the four-legged projectile hit Bob full force and knocked him to the ground. Owner and dog rolled over and over with much laughter and licking.

"That's some dog you have there," observed one of the picnickers.

"Oh, he's a lover," agreed Bob, rubbing Shaggy's ears fondly.

"You mean a killer," muttered a fearful woman at the back of the group who tightly clutched the hands of two children straining to get closer to the dog. "He'd better keep that dog on a leash or I'll report him to the park ranger!"

The Espositos took great pride in their devoted Doberman, Madd Max, because he defended their property so heroically. After watching Max romp with the rottweiler next door, little

Joey invented a new game to play with his dog: He would hide behind the couch, wait for Max to walk by, then growl ferociously, lunge at the dog, and try to pin him to the floor. The first time Joey tried his new game, Max whipped around and lashed out at the child, who toppled backward. Screaming, Joey clutched the side of his face.

"Max is vicious; I want him destroyed," Rose Esposito insisted when she brought Joey home from the emergency room.

"No, Mommy! It was my fault. I scared him," sobbed Joey. "Please don't kill my dog!"

Al Esposito looked at the fourteen stitches in his son's swollen and bruised face. Then he looked at Max lying at Joey's feet and staring at the boy with that familiar look of undying devotion.

Every time the Bennetts' doorbell rings, their springer spaniel, Photon, performs a dancing, barking ritual that led one guest to describe her aptly as "a mindless yapping cyclone."

"She's so devoted to us," the Bennetts invariably apologize to startled visitors. "We can't seem to convince her we don't need all this protection."

When the local dog officer served the Bennetts with a warning citing Photon as a public nuisance and threatening court action if the behavior persisted, the family felt crushed. At least one of their neighbors, sharing their guest's view of the yapping dog, had filed a complaint. The Bennetts ricocheted between embarrassment, guilt, remorse; and anger aimed at themselves, Photon, and their neighbors. These varied and conflicting emotions proved to be such a drain that the Bennetts sold their suburban dream home and moved to the country.

In this book we're going to explore how body language and its attendant emotions affect our relationships with our dogs. A few years ago the idea of body language as an important form of nonverbal communication preoccupied many psychologists and writers, who told us how to interpret the signals lovers, spouses, and bosses were giving us with their bodies. Armed with our lists of body signals and their associated meanings, we proceeded to "read" other people, thus reducing their need for verbal expression of conscious (and unconscious) emotions. Unsuspecting women who crossed their ankles would find themselves branded brazen hussies by other women while simultaneously deflecting the unwanted attention of men responding to their silent come-hither calls. Harried businessmen who loosened their ties and ran a hand through their hair unwittingly revealed to all those privy to the vocabulary of body language that they were insensitive and too attached to their mothers. Eventually the fad died out as it became clear that interpretations of body language are so subjective that any signal sender and any receiver could attach totally different meanings to any gesture. When Shiela assumed that the way her boss sat during meetings indicated deep-seated insecurities and latent sexual problems, you can imagine her embarrassment when she learned that he sits that way because he's recovering from a total hip replacement.

I asked a psychologist friend to help me understand the relationship between body language and emotion in people. From my own experience I knew that a physical signal such as a wink could mean one thing to one person and something entirely different to another. But weren't there some general rules of interpretation?

"Yes—sort of," my friend said. "However, the meaning of physical cues goes far beyond simple cause and effect. If I winked at you during a boring lecture, you'd interpret my gesture as our sharing an inside joke; but if you winked at a strange man in a cocktail lounge, he'd interpret your cue as a sexual invitation. Still, even though the meaning of the signal depends on the situation and on the relationship of the people involved, we can loosely categorize signals. For example, a person standing at the side of the road with his thumb in the air wants a ride."

We explored this example further, discussing how the hitchhiker's stance could trigger a wide range of emotional responses in passing motorists. Shoulders back, eyes glaring, feet planted firmly might suggest that this person wouldn't be a safe passenger, whereas a bright but weary smile might evoke compassion.

"It's a two-way street," my friend continued. "Body language may express a variety of emotions, and the interpreter of the signals adds his or her own emotions to the process. A driver who's been mugged and robbed by a hitchhiker obviously views all hitchhikers differently than one who's only had positive experiences picking up riders."

"I think I've got it," I said. "If the IRS audits ten percent of all taxpayers, that's body language; if they audit me, that's emotion. If the fourth-grade band performs with distinction in the statewide competition, that's body language; if my son's playing the tuba, that's emotion."

My friend laughed. "If a beautiful woman smiles and waves at everyone in a crowd, that's body language. If she smiles and waves at me, that's emotion."

From this exchange we can see that it's impossible to separate body language from the specific situation and the emotional condition of either the sender or the recipient of the physical cues. This also holds true in our relationships with our dogs. Shaggy's exuberant body language expressed love to Bob Kalish and was a sign of intelligence to one picnicker; however, it triggered fright in the woman who fears unleashed dogs. Photon's barking body language conveyed a message of misguided protection to the Bennetts, but that same frenzied barking meant nothing but a noisy nuisance to their neighbors. Madd Max? Little Joey inadvertently sent the Doberman body language messages that demanded a response far different from the one the child intended.

SETTING THE STAGE

I've never forgotten something Pastor Tejan told us one day in Sunday school class when we were discussing what sort of behavior it took to get to heaven. The pastor insisted we'd have to do more than go to church every Sunday and say our prayers each night. "There's nothing wrong with simply being good," he concluded. "But you have to be good for something. Otherwise you're good for nothing."

Many years later I remembered these words when I came

across an incident related by William Campbell in his marvelous book *Behavior Problems in Dogs* (American Veterinary Publications, 1975). When Campbell needed some photographs of dogs exhibiting aggressive biting behavior, a local guard-dog trainer let him use one of his dogs as a subject. Presenting the leashed shepherd, the trainer gave the command, "Watch it!" which he always used to order the dog into a fully aggressive attack stance: body rigid and poised to spring, ears erect, teeth bared, all accompanied by loud growls and snarls. Surprisingly, the dog remained indifferent even after Campbell deliberately tried to provoke it. While handler and would-be photographer wrestled with the problem, the dog lifted his leg on a nearby bush, then turned and growled menacingly. It wasn't that the dog had forgotten his training or chose to ignore it; he merely wanted to be aggressive "for something." Once he'd marked his territory, he was quite willing to display the necessary body language to protect that territory.

As we embark on our quest to interpret the exchange of body language and emotion between humans and dogs, we should bear in mind that any such exchange must serve some purpose for both participants. Because dogs use various postures to indicate their relationships to their environment, we need to recognize them as valuable clues about their emotional life; and we must not only recognize the signals, we must learn to assign appropriate meanings to them. Otherwise we'll never get the most from our relationships with our dogs. Think of Al Esposito watching Max stare adoringly at Joey. Before the biting incident the Doberman's expression meant only one thing to Al: devotion. Joey still sees love in that expression, but Rose now interprets it as evil and predatory. Al isn't so sure. Is Max worried? Devoted? Evil? Or does the Doberman feel anything at all?

An understanding of human/canine body language and emotion must somehow fuse two seemingly antagonistic schools of thought:

- The unemotional, highly objective approach of the animal behaviorists (ethologists).

- The emotional and highly subjective view of the average pet owner.

It would be quite easy for Cupcake McCorkle's owners to insist that scientists who study wolves and wild dogs don't know anything about miniature poodles. Similarly, more than one behaviorist has openly scoffed at owners who believe their dogs respond to them purely out of love and devotion. However, such conflicting views do little to expand a curious owner's knowledge of canine psychology and may actually hinder anyone seeking to create the best possible relationship with a pet.

Let's take a look at some of the differing beliefs of objective behaviorists and subjective owners that have the most influence on dog/owner relations.

DOMINANCE AND SUBMISSION

Scientists who studied wild dogs and other canids (foxes, wolves, coyotes) quickly classified two basic types of body language animals exhibit when interacting with one another: dominant and submissive behaviors. Dominant body-language displays typically include:

- Placing the chin on the other's shoulders.
- Growling or snarling if the other whines or attempts to move.
- Placing the front paws on the other's back.
- Circling and sniffing the other.
- Holding the ears erect.
- Holding the tail erect.
- Looking directly at the other.
- Deliberately marking the area with urine.

Submissive behaviors include:

- Tail lowered and even curled under the body.
- Ears flattened against the head.
- Gaze averted.

- Rolling over and exposing the abdomen.
- "Nervous" licking or swallowing.
- Cringing or trembling.
- Seemingly involuntary dribbling of urine.

Although other dominant and submissive signals do exist, the major ones listed above basically describe the body language wild dogs use in their interactions with other animals.

Behaviorists don't make value judgments about dominance and submission but simply define them as *unemotional states of being.* Dominance and submission aren't good or bad traits, just forms of behavior that serve to keep the peace whenever two or more animals relate. As long as one animal responds more dominantly to the other's more submissive display, a fight will not occur. In nature this recognition conserves critical energy stores; when life depends on seeking prey, mating, or caring for young, constant squabbles over rank would threaten the survival of the entire species.

From the behaviorists' point of view, these often highly involved and intricate displays have resulted from evolutionary trial and error. When wild dog A curls his lip, growls, and walks stiffly toward dog B, the behaviorist insists that A doesn't think, "I'm more qualified to lead this pack then B, so I'll show him who's boss!" but merely responds to a set of unemotional instinctive nonthought patterns that lie as much beyond his awareness and control as his eye or coat color.

"Hogwash!" screams every dog lover in the universe. "My dog not only thinks, she loves and gets mad and feels all the emotions I do." Regardless of your own opinion, don't summarily dismiss the behaviorists' view; it does offer insights for all dog owners. If we occasionally mimic the scientists, divorcing emotion from canine behavior in an effort to understand it better, we become like the entomologist who uses black and white films to study butterfly flight patterns without being distracted by the subjects' vivid colors. Such approaches produce information that can be extremely helpful and instructive, even though they rely on assumptions that take the color out of the picture. Eventually, information about butterfly flight increases our appreciation of the full-colored form; and eventually, unemotional studies of dominant and submissive behavior can broaden our understanding of those happy-go-lucky canines that have snuggled their way into our hearts.

SPIKING THE BEHAVIORISTS' PUNCH

Let's see how two typical dog owners would interpret our list of dominant and submissive body-language displays. Our first owner, Emily Sullivan, is a systems analyst who owns a male Scottish terrier, Brittamark's Bit O'Honey of Skye. Our second owner is Lou Rutherford, a physical-education teacher, football coach, and proud master of Merlin, a black Labrador retriever. Compare the way Emily and Lou interpret the behaviorists' findings!

Behavior	Emily	Lou
Placing chin on shoulders	"He's expressing love."	"He's showing who's boss."
Growling or snarling if other moves	"He's saying, 'Don't leave me.' "	"He's saying, 'Don't move!' "
Placing paws on other's back	"He wants to make love."	"He wants the other to lie down."
Circling and sniffing	"There can't possibly be any meaning for anything so disgusting."	"He's checking him out."
Ears erect	"He's listening."	"He's paying attention."
Tail erect	"He wants to play."	"He's ready for action."
Looks directly at other	"He trusts you."	"He's watching very carefully."
Marks with urine	"He's being spiteful."	"He's challenging you."
Tail lowered	"He's scared."	"He's a coward."
Ears flattened	"There's too much noise."	"He's a coward."
Averted gaze	"He's shy."	"He's shifty, guilty about something."
Rolls over and exposes abdomen	"He wants to be loved."	"He's a chicken."
Cringes or trembles	"He's scared or cold."	"He's chicken."
Dribbles urine	"He's scared."	"He's being spiteful."

As you would expect, neither Emily nor Lou perceives canine body language the same way the behaviorists do. Actions objective scientists view as totally void of any thought or emotion, subjective owners almost always see as the result of *both.* Furthermore, the two owners may assign completely different and, in some cases, quite opposite meanings to identical behaviors or body language. Not only do their perceptions of the emotion underlying the dog's behavior differ, their own emotional response to that body language differs as well. Because Emily believes a lowered or tucked tail signals fear, she immediately feels sympathy for any dog displaying that behavior. On the other hand, Lou perceives this same behavior as a sign of cowardice and is repulsed. The behaviorists point to this obvious lack of agreement as proof that because emotion is "in the eye of the beholder," it doesn't exist in dogs at all.

Such inconsistencies and their implications bring to mind the story of the census taker who cautiously approached the elderly New Englander sitting on her porch with her growling hound dog at her side.

"That dog bite strangers?" asks the nervous public servant.

"Nope," says the old woman, where upon the census taker enters the yard and is painfully nipped by the dog.

"I thought you said your dog doesn't bite strangers!" the man roars.

"He don't—only census takers."

People unfamiliar with the Rhodesian Ridgeback breed experience a more pleasant surprise. The characteristic raised ridge of fur along the spine of these dogs causes many strangers to assume they have encountered a frightened or angry animal. What a relief to learn that this ever-present characteristic carries no more behavioral or emotional significance than eye or coat color!

If the interpretations of body language span such a broad spectrum, and if the signs themselves can vary from breed to breed and from person to person, how can we possibly use body language and emotion to enhance our relationships with our pets?

CONSISTENCY

In *Through the Looking Glass* the caterpillar responds to Alice's criticism of one of his rather peculiar definitions by saying that words mean precisely what he wants them to mean. As long as he uses the same word to mean the same thing, it doesn't matter what it means to others. They'll eventually figure it out. Although this sounded ridiculous to Alice and comical to generations of readers, we dog owners adopt the caterpillar's position all the time. When Bit O'Honey stares at Emily Sullivan and Emily responds, "Is it time for baby's ice cream and cookies?" the owner has attached a highly complex and emotional meaning to what an impartial observer might consider an almost insignificant body-language signal. When Merlin stares at Lou Rutherford, and Lou tells his poker pals, "Gotta let Merlin out. He wants to chase some rabbits and show the neighbor's mastiff who's boss," he's tacked an equally complex but totally different meaning on to the same behavior.

We can take a cue from the dispassionate behaviorists here, avoiding judgment of Emily's or Lou's differing interpretations as right or wrong and examining how they affect the specific owner/dog relationship instead. Given Emily's response to Bit O'Honey's staring, the dog will probably dash madly to the refrigerator to wait for Emily to produce the promised treats. If his owner doesn't fetch the treats, he'll repeat the stare-and-dash routine until she does. Similarly, Merlin may bring home a dead rabbit and display some bloody puncture wounds his owner will attribute to battle with the mastiff next door.

Regardless whether the stares are objectively related to Bit O'Honey's passion for ice cream or Merlin's desire for combat, the owners think they are. Again, most animal behaviorists would dismiss such associations as arbitrary and groundless. However, as the caterpillar knew, such associations do work for their creators.

Therefore, one of the first skills we learn consciously or unconsciously to use when we interpret our own dogs' body language is consistency. Conversely, our own consistency helps our dogs read *our* body language, too. If Emily always produces ice cream and cookies whenever Bit O'Honey stares at her in a particular way, he'll inevitably connect that behavior with her response. If the cookies and ice cream didn't always appear when Bit O'Honey flashed the signal, he would probably make no connection and abandon the behavior.

When dog and owner respond to each other's body language consistently, the definition or meaning of that body language becomes quite predictable and real to both. Doesn't this mean our dogs merely respond randomly and *we* supply the emotion and meaning? Behavioral studies show that animals instinctively respond to a dominant presence or to fear in one of three ways: they freeze, they fight, or they flee. Let's consider a typical frozen dog response to an owner who shouts, "No!" In the owner's mind the dog has responded to a command *and* understands the complete message:

"No!" (Don't leave the yard.)
"No!" (Put down that shoe.)
"No!" (Don't chew the rug.)
"No!" (Don't bark at my mother.)

The behaviorist analyzing this same situation says that the dog has responded to the owner's *position;* it responds to the owner's "No!" the same way it would to the growl or snarl from a dog it perceives as more dominant. It's not that it's stopping any particular (negative) activity: it's stopping *all* activity in response to the owner's dominant rank and the freeze/fear response it evokes.

Should we accept the behaviorists' view that our dogs don't respond to us emotionally, that we only pretend they do? Although behaviorists believe that scientific validity can only be maintained by eliminating anything as subjective and variable as emotion from interspecies interaction, few dog owners would keep a pet if it weren't a source of emotional exchange. Even if we accept the notion that we create all the emotion that surrounds our relationships with our pets, we can improve these relationships by determining whether our interpretations of our dogs' body signals are consistent and beneficial for us and our pets.

Unfortunately we often unknowingly reinforce certain body

language and meaning linkups in our dogs, but when we come to perceive the dog's behavior as negative, we throw up our hands and say, "I just don't understand why dogs behave the way they do!" So the Bennetts unwittingly encouraged Photon's doorbell-related barking and carrying on because they thought it was cute when she was a pup and because the behavior fulfilled their definition of normal protective dog behavior. Their consistent acceptance and even approval firmly reinforced the behavior. When the police warning revealed that their neighbors didn't agree with their definition of the body language, the Bennetts suddenly felt betrayed—and it was easier for them to believe they'd been betrayed by their neighbors than by their dog and their beliefs.

While consistency and benefit/liability may appear to be quite unrelated criteria to apply to our interactions with our dogs, if we consistently link body language to emotion in our own behavior and that of our dogs, we inevitably create what we consider beneficial or detrimental results. If the behavior didn't hold any meaning for us, we wouldn't consistently link it with anything and would therefore ignore it. If we believe a particular behavior holds meaning, we often assign it a meaning that makes sense to us. Emily wouldn't dream of turning Bit O'Honey loose to hunt rabbits or attack mastiffs any more than Lou would feed Merlin cookies and ice cream. Although each thinks the meaning of the dog's stare originates with the animal, the fact that the owners fulfill their roles (dishing up ice cream, opening the door) means they find this interpretation acceptable, even beneficial. If they didn't, they would never respond to the dog's body language long enough to ensure its incorporation into the relationship. Furthermore, if the dog didn't find the owner's interpretation acceptable, it wouldn't consistently participate. If Bit O'Honey refused to eat ice cream and cookies, Emily would try to figure out what else her dog was trying to communicate with his stare. If Merlin sat on the porch and howled every time Lou put him out, Lou could easily decide the stare meant the dog wanted something *inside* the house.

All would be fine as long as we could guarantee that our interspecies interpretations were physically, mentally, and emotionally beneficial to owner and dog. Unfortunately that is not always the case. For example, when Bit O'Honey's stare-related behavior leads to obesity and Merlin winds up crippled for life after a particularly vicious battle with his canine foe, we must question the value of the body language and emotional connections for the dogs. When body language and emotion linkups lead to the kinds of mental anguish the Espositos and Bennetts experienced, to say nothing of the physical pain Joey experienced, we can hardly call these relationships beneficial for either party.

THE CANINE GALATEA

The idea that we can enhance or change various aspects of our relationships with our dogs through our awareness and use of body language and emotion makes owning a dog a tremendously intriguing experience. If our dogs were simply beagle-shaped bundles of genetically programmed behaviors and responses, we wouldn't have nearly as much fun with them. Anyone who's watched a golden-retriever pup instinctively retrieve a ball can logically attribute that activity to the dog's instincts and breeding. However, the pleased grin and the proudly waving tail elevate the mechanical act to a high personal plane. My dog goes belly-up for lots of reasons, all of which the behaviorists would classify as submissive. When she goes belly-up beside me on the blanket under the tree with her wide-open eyes staring deep into mine and her nose a scant half inch from my sandwich, that's a very special form of submission indeed! Setters were bred for their ability to move along the ground on their elbows to facilitate stalking birds, and that result can be viewed as the culmination of careful genetic manipulation. When English setter Molly gently "stalks" her three-year-old master around the living room and uses her front paws more like long silky hands to steer him away from the stairs, the result can only be described as beautiful. However, we're not just talking about the beauty each owner beholds in his or her own dog. We're talking about something far more creative.

Greek mythology tells the tale of Pygmalion, king of Cyprus, who wanted to carve a statue representing the most perfect woman in the world. Pygmalion succeeded, but his creation was so lovely he fell head over heels in love with it. Had the goddess Aphrodite not given life to the beautiful Galatea, Pygmalion would have suffered a rather empty, miserable life.

There's a bit of Pygmalion in every owner and bit of Galatea in every dog. We meet that young pup for the first time and envision a lifetime of companionship and happy experiences stretching out before us. But our dogs aren't lifeless lumps of clay; they respond to each impression we make on, with, or against them, and the final form of our creation may turn out to be something far different from what we had initially expected. Furthermore, while we're concentrating on how they're responding to us, we often ignore the fact that we're responding to them. Thus in our role of Pygmalion creating the perfect companion, we may not only *not* get what we want, we may ourselves become soiled, exhausted, and quite disheartened in the process, longing for an Aphrodite to bail us out.

Fortunately, although each owner responds emotionally in a unique way to his or her dog's body language, the majority of us fall into one of three classifications in any given situation. Recognizing your own orientation will help you take the next step toward enhancing the body language and emotion that benefits your relationship with your dog and eliminating those that undermine it.

MOMMY'S LITTLE BABY, DADDY'S LITTLE GIRL

Some owners equate owning a dog with "owning" a baby or other dependent human being. Studies have indicated that pugs, Boston terriers, Pekingese, and similar large-eyed bracephalic (squashed-nosed) breeds appeal to some people precisely because they look so much like infants. Other studies suggest that some people, fearful of bearing or unable to bear children, use dogs as surrogate offspring, outlets for their unfulfilled parental needs. Still others say that many people relate to their dogs as people because they find it difficult to relate intimately to other humans.

In many situations, owners will quite naturally view their pets as furry little humanoids simply because they try to relate dog behavior to what they know best: human behavior. People who have had no or limited experience with pets fall easily into the trap of relating to dogs as little people in fur coats. Imagine a Pygmalion who never created anything but human statues, whose patron demands he carve a Shih Tzu. If Pygmalion knows little about dogs in general and nothing about Shih Tzus in particular, he must rely on his knowledge of human anatomy and only hearsay evidence of the dog's. Under these circumstances the result could easily look more human than canine, or something dreadfully in between. The latter calamity befalls owners who interpret their dogs' body language in strictly human terms and expect their dogs to respond in kind. Owners who reinforce and encourage dominant behavior, interpreting it as protective, can be shattered when their dog "protects" them from a beloved visiting grandson or neighborhood child.

While women may express this kind of orientation more openly, many men fall victims to it, too. Bit O'Honey is Charlie Sullivan's little boy; as such he shares his master's favorite foods and gets an extra blanket when Charlie himself feels cold. Like the rest of the Sullivans, Bit O'Honey participates in all family rituals and celebrations. If a Sullivan receives bad news, he or she shares it with the Scottie, whom they expect to be particularly attentive and solicitous. Similarly, if the family believes something unfortunate has befallen their pet, they shower attention on him until they're convinced he's feeling better.

One area where this dependent-species orientation can have great value is in pet therapy. Emotionally troubled children and adults who can't relate to other people can often relate to pets. By projecting human characteristics on animals, they can deal with behavior and situations that would be threatening if they originated with another person. Similarly, many depressed or lonely people have found that indulging their pets in human terms effectively distracts them from their own problems. In such ways pets can ease humans down a wide variety of obstacle-ridden paths. However, when dogs are used to facilitate human health, dog and person must be matched with the utmost care. Therapists trained in this art (and it is more art than science) strive to balance those projections that speed human recovery against those that may be detrimental to the dog.

In everyday life the dependent-dog orientation works fairly well because owners who baby their dogs always know that their dogs are thinking and feeling; all they need do is evaluate their own thoughts and feelings. This works fine until the dog does something "inhuman." For example, when Bit O'Honey joyfully munches caked horse manure off the Sullivan daughter's riding boot rather than Emily's homemade double-chocolate-chip cookies, Emily flounders in a sea of unpleasant emotions: repulsion, anger, disgust, guilt, remorse. Because she views Bit O'Honey as a furry humanoid, she can't accept manure eating as normal activity, even though as far as Bit O'Honey's concerned, it is.

Owners who view their pets as dependent humanoids also have a tough time coping with their dogs' normal responses to other animals. Emily already told us she finds the sniffing and circling behavior of dominant dogs disgusting and territorial marking spiteful. Were her husband to circle and sniff her when she returned home from the supermarket, were he to race into the front yard to urinate on every tree whenever a stranger passed, she could appropriately label these behaviors "disgusting" and "spiteful," to put it mildly. However, such common, normal, and natural canine behaviors shouldn't be judged by inappropriate human standards; doing so can only undermine the relationship.

One of the saddest incidents I ever experienced involved a couple who owned two male littermate pups. Because the owners firmly believed in peace and nonviolence, any time the pups engaged in what the owners considered rough behavior, they separated the dogs. Unfortunately, while adhering to their personal convictions, they denied their pets the opportunity to establish a pack structure in their natural, safe, and harmless way. Whereas other pups worked out a hierarchy via play-fighting, tumbling, and baby-teeth nips, these two were admonished to "play nice" and "be gentle" and were separated whenever they didn't comply. As a result the tension between the dogs grew. They developed more powerful permanent teeth and replaced puppy games with aggressive adult displays. In the owners' misguided effort to ensure a loving, peaceful relationship between their pets, they virtually guaranteed the opposite.

One hot August afternoon shortly before the dogs' first birthday, the owners chained one of the normally untethered animals because the dog seemed a little under the weather. When the owners went into the house, a savage battle erupted in the yard. In the short time it took them to race outside in response to the ferocious snarls and screams, the hierarchy had been established: One dog, still tethered, lay dead; the other was badly hurt.

I'll never forget the anguished look on the wife's face as she told me what had happened. Having devoted her whole life to the belief that violence in any form was avoidable and unnecessary, she was shattered. My explanation that what her pups had tried to express, what they wanted and needed to express, wasn't violence at all, did little to reassure her. "Then it's as if I killed that dog myself!" she cried. And as much as I wanted to console her, it was impossible for me to disagree. By imposing her personal definition of violence on a normal expression of canine body language that had nothing to do with violence at all, she had in fact sealed her dogs' fate. Not only had she lost one of her beloved pets, her confidence in herself as a pet owner had been devastated, and she was overwhelmed by guilt. Unlike Emily, who was able to confront and dispell her emotions in a relatively short time, these owners saw the consequences of their own beliefs every time they had any contact with the surviving dog. They found it so impossible to relate to the animal that they had it euthanized within the month.

From this discussion we can see that viewing our dogs as little people in fur suits offers advantages and disadvantages. As we all know, sometimes it's a distinct advantage to be able to put ourselves in another person's place. This is particularly beneficial following some negative interaction. We *know* we feel terrible, and by assuming the other does too, we're more apt to try to reestablish a stable relationship as quickly as possible. On the other hand, overlaying the other's behavior with "I know what *I'd* feel and do in this situation"—and judging the other accordingly—can be problematic. Whenever we're tempted to view our dogs' behavior in this light, we owe it to ourselves and them to remember *we are different species with different needs.* While it's true we can share similar experiences, the chance of our dogs and us having the same interpretations of these events is nil. For example, when we take our often highly subjective and emotional views of birth, death, and diet and apply them to problems our dogs are experiencing in these areas, we can create unnecessary grief for ourselves and our pets. We all know owners who allow their pets to flounder through high-risk pregnancies in order to produce unwanted puppies because they find abortion or spaying barbaric. On any given day countless dogs are kept alive via extraordinary technological and medical feats—not because there's any hope that the animals can ever live quality lives, but because their owners are afraid of death. How many canines waddle about bulkily or can barely maintain themselves in their appalling nutritional states simply because their *owners* find balanced canine rations distasteful and unappetizing?

LOBO IN SUBURBIA

While some owners, lacking experience or confidence in their relationships with their dogs or the existing situation, tend to relate to their pets as people, others relate to them as wild animals. Although this phenomenon can develop between any owner and any breed, it seems to occur most frequently between men and large breeds, especially German shepherds, Dobermans, and so-called primitive northern breeds, such as huskies and malamutes. This may simply be because the latter's more wolflike appearance facilitates the association, just as the brachycephalics' infantile appearance makes them easier for some people to baby.

If owners who adhere to this orientation attribute nothing more or less to their pets than the unemotional instinctive responses described by the behaviorists, they probably won't encounter many problems. Although the resulting relationship might strike some dog lovers as rather sterile and mechanical, the dog's behavior should seldom surprise the owner.

However, as we noted earlier, emotion spices pet ownership, and practitioners of the "wild dog" philosophy flavor their relationship with their own special brand of emotion. Whereas those who view their dogs as dependent, fur-coated humanoids act like doting parents, treating their pets as cuddly and potentially well-behaved family members, those who view their dogs as displaced wild animals see themselves locked in battle with their pets, struggling to stay one step ahead of, and be top dog over, a canine eager to topple them the instant they let down their guard. Some owners even go so far as to mimic dominant dog behavior to ensure their rank. (One husky breeder joins nightly howling sessions with his pack and clamps each dog's muzzle in his own mouth to signal his dominance.)

Don't think owners who use this approach dislike their pets; they do enjoy them, but they derive much of their enjoyment of the interaction from their ability to dominate the animals. One shepherd owner persisted in loud shouting (growling and snarling) and heavy-handedness (dominant pressure on the dog's back and shoulders) long after the dog had become perfectly well behaved. He's one of those owners who read discussions of dominant and submissive body language in wild dogs and erroneously interpreted them to mean

"Get them before they get you"
"Kill or be killed"
"To the victor belong the spoils"
"Survival of the fittest"

or similar sentiments erroneously defining dominance as winning some battle. Even the behaviorists would say that such beliefs project human values on to dogs. The shepherd spent most of his life being yelled at and jerked to come, sit, stay, and heel for no reason other than to fulfill his owner's need to display dominant behavior. It never dawned on the owner that the dog would have willingly obeyed nothing more "powerful" than a soft voice command or even a subtle hand signal.

In such a relationship it's difficult to believe that the owner's goal is obedience. In the case of the dominated shepherd, the owner really wanted a fight, but the dog never obliged him— showing a degree of patience that surely entitles him to canine sainthood. When the shepherd died, the owner wasn't sure whether he should get another shepherd or try another breed. "A man with my tremendous energy level needs a dog with a lot of spirit," he said. "That other dog never showed much spunk."

As with the anthropomorphic view, the issue isn't whether the wild-dog orientation is right or wrong, but whether such an orientation allows for beneficial interpretation and exchange of body language and emotion. If the dependent-species view distorts the relationship because the dog can't develop its own unique species identity, the wild-dog view can equally hamstring a relationship by creating too much distance between owner and pet.

For example, some owners who view their dogs this way can abandon or euthanize one dog and get another with less thought than most people give to replacing a light bulb. In that case, the distance between owner and dog doesn't even acknowledge that both species are living beings. While dealing with owners who treat their pets like babies can create nightmares for veterinarians, dealing with the wolf-dog owners who demand euthanasia for little more reason than the way a dog looks can be downright agonizing.

Unfortunately, it always seems to take some horrible incident to point out the limitations of this orientation. Adam Christensen believed his dog, Apollo, should run free and be able to do "dog" things. One of the natural activities Apollo chose to pursue was his instinct to chase and sometimes attack moving objects. In this case, the moving objects were three of Adam's neighbor's prized sheep. The neighbor drove into Adam's yard with Apollo, the dead sheep, and his .22 in his pickup.

"You want to do it, or do you want me to?"

Adam truly loved his dog, and the memory of the dog's trusting expression up to the instant he pulled the trigger will never leave him. Such a brutal example may be relegated to rural areas, yet thousands of dogs exercising their natural chase instincts are killed by vehicles every day. The owners didn't pull any triggers, but they wind up feeling as if they had.

Whenever we find ourselves in situations where our response to Tippy's behavior is "It's him or me" or "He's on his own," chances are we feel somehow threatened by the interaction and wish to distance ourselves, either from the consequences of our own behavior or from the dog itself. In this case we owe it to ourselves and our dogs to view the experience as shared and co-created rather than totally alien and incomprehensible.

THE BONDED VIEW

Somewhere between Cuddles the baby and Satan the wolf-dog lies the ideal owner/dog problem-solving orientation, one that takes into account the fact that dogs and people have their own special needs and body language. Whereas the Sullivans are constantly looking for ways Bit O'Honey acts like a person, and whereas Adam Christensen was captivated by Apollo because he was so wild, bonded owners relate to their pets as domestic dogs. They view dogs as a previously wild species that has been physiologically and behaviorally tempered by thousands of years of intimate human contact. This contact doesn't make dogs human any more then the countless varieties of hybrid corn are more human than wild maize. The human interaction simply provides a different *environment* in which both species can exercise their innate or instinctive behaviors in different ways. Sometimes these are compatible, sometimes not. However, the bonded view maintains that whenever canine and human needs and interpretations do coincide, the results are no more "right" than those times when they don't. Owners and dogs who form strong bonds accept the fact that their differences hold them together as much as their similarities.

Think about the characteristics that attract you to your dog. My list for one of my dogs looks like this:

- Keeps me company.
- Is always happy to see me.
- Is even-tempered and gentle.
- Is quiet.
- Is generally well behaved.

Now let's see a list of not-so-good qualities:

- Goes into the swamp.
- Acts uneasy around strangers.
- Rolls over submissively at the least provocation.
- Is not particular about her personal hygiene.

If your dog could list your most redeeming qualities, what would they be? My dog would probably say I'm:

- Affectionate and talkative.
- Always happy to see her.
- Gentle.
- Generally (predictably, consistently) well behaved.

My not-so-good qualities would most likely include:

- Goes away for a whole day or even more.
- Allows strangers on the property.
- Is too fussy about flower and vegetable gardens.
- Growls unnecessarily, particularly at certain times of the month.

As you look over my lists and your own, you see areas where we owners and our dogs agree and areas where we disagree. Contemplate your areas of disagreement. Do they bother you? Do you believe they bother your dog? For example, I know my dog doesn't like strangers and would prefer that none ever entered our yard. However, I have no intention of prohibiting visitors, nor do I feel guilty about my dog's reaction when they arrive. Similarly, I'd prefer she didn't come back from the swamp with that "Clean me!" look in her eye or roll over and pee submissively if I sneeze loudly. However, she displays both these forms of behavior, and I've learned to accept them. If I have time, I brush her and enjoy it; if I don't have time, she stays ragged and bedraggled for another day, and I don't give it a second thought. I recognize her submissive behavior as exactly that; it doesn't mean she's a coward or that I or someone else have abused her. If it made me feel sad or guilty, or if I believed it made her unhappy or unhealthy, I'd spend the necessary time to change the behavior. I'm content with our relationship just the way it is; and I believe she enjoys it too.

I'm neither her mother nor her master. She's not my baby; unlike other dogs I've had in the past, she doesn't sleep on the bed and rarely wants to come indoors. Nor is she a wild creature. She enjoys sleeping on her braided rug by the door, eating her regular and quite unwild commercial-dog-food meals, and sitting on the picnic table surrounded by classical music and surveying her domain while I write. We're good friends as only members of two different species can be, each drawn to the other as much by curiosity as anything else. "What in the world are you up to now?" her eyes ask as she watches me plant the garden. "You crazy fool, get out of there!" I shriek when I spy her rolling with obvious canine sensual delight in the newly turned earth a half hour later.

But in addition to our differences, there's a certain inexplicable kinship that binds us together. Having tried to define it objectively, I can see why scientists find it much easier to ignore it or deny its existence. Can we call it love? Absolutely, but liberally sprinkled with mutual patience. What binds me to her and all my dogs more than anything is a sense of awe, an awareness that this animal enables me to experience and share a unique form of communication unavailable with any other living thing. No other dog can experience the "me" I share with her; no other human receives from her what I do. Just being aware of that makes me want to make that exchange as beautiful and mutually beneficial as possible.

What I've just described comprises what I consider to be the best relationship with my dog. I've experienced it and I know it's very real; and I use it as a beacon when problems arise between us. When I catch myself reading her expression as mournful, condemning, or heartbroken as I pull out of the driveway, I resist the urge to equate her behavior with the memory of my son's face when I dropped him off for his first day at a new school. "She's a dog, not a child," I remind myself. When she comes out of the swamp caked with something chokingly foul and sticky, I resist the urge to banish her. Had she either the hair coat or the swamp-savvy of a wild creature, this wouldn't have happened; I know she's uncomfortable, I know she needs my help.

The bonded approach to our pets and any problems simply requires that we remember we're in this together. Our dogs aren't people, nor are they alien, wild beasts. Our dogs are dogs and want a good relationship with us just as we want a good relationship with them.

In the next chapter we're going to survey briefly the basic body language we associate with the most common emotions that give our relationships with our pets their distinctive flavor. Does your dog give you consistent clues that let you know she's happy? Do you return the favor—or do you expect her to read your mind?

2
READING
THE SIGNS

Normally when Joe Spivak delivers the mail to the Watermans, he carefully surveys the yard for their nasty cocker spaniel, Pumpkin, before unlatching the gate. Over the years Joe has learned the hard way that an ounce of caution is worth a pint of iodine and a pound of pain. Once when it seemed the dog lay sleeping in a hammock on the porch, he pounced on Joe's back just as Joe leaned down to slip the mail through the shiny brass slot. More recently, Pumpkin had launched an attack from behind the lilac bushes, and just last week from inside what looked to Joe like an empty box.

Today, however, the mailman was running late and had blithely strolled halfway up the Watermans' front walk before he realized he'd made a grave tactical blunder. Pumpkin stood on guard on the front steps, and there was no evidence the Watermans were at home. There stood Joe, halfway between the front gate and the house, nervously shifting his mailbag from shoulder to shoulder. Didn't Pumpkin look exceptionally nasty today? His fur bristled, and he held his body rigidly, staring eyes fully dilated and evilly black, lips curled back, glistening white fangs exposed. Menacing growls reverberated through the hot July air. Joe looked around frantically for a weapon.

Was the dog actually growing bigger or had he crept closer? Joe's eyes darted from the dog to the front gate, which now seemed a mile away. Holding his bag in front of him like a shield, Joe cautiously began backing up as Pumpkin took one stiff, ominous step forward. Suddenly Joe's foot landed on something hard and round, and he almost lost his balance; it was a solid rubber ball.

"By God," thought Joe, snatching the ball. "It's him or me!" In one swift motion he threw the ball as hard as he could, but it sailed harmlessly over Pumpkin's head. Joe braced himself for the attack. Pumpkin watched the ball fly past, then cocked a malevolent eye at the mailman. Before Joe could move, the dog streaked by him, retrieved the ball, and dropped it at Joe's feet, tail wagging joyfully. Half-dazed, Joe picked up the ball and threw it again. And again. From that day until his retirement, Joe always carried a ball in his bag.

Although I've never seen any studies on ball-dependent body language and emotion in dogs, Joe's story didn't surprise me. I had a delightful patient (also a cocker) who would permit any sort of handling as long as he clutched his tennis ball in his mouth. At first I felt awkward inserting a rectal thermometer into an unrestrained dog standing on a slippery stainless-steel table while the owner lounged casually against the wall murmuring encouragement to his pet. What if the dog suddenly jumped off the table and hurt himself? What if he dropped the ball and hurt *me*? But the owner had so much confidence in and was so strongly bonded to his pet that I soon put my fears aside and learned a valuable lesson from the incident.

In both these examples it would take a computer to analyze all the body language and interpretations between human and dog that culminated in such unusual episodes. But regardless of the process, the interactions become quite tolerable once human and dog agree on the meaning of the relevant body language.

These two examples demonstrate once again that any interaction between human and dog contains several components:

- The dog's body language.
- What the body language means to the dog.

- What the body language means to the owner or other people.
- How the owner or other people respond or relate to the behavior and its associated meaning and emotion.

And although we can't prove it, most of us function as though our dogs perceive interactions with us in a similar fashion. My own dog's breakdown of her interactions with me might look like this:

- What my owner is doing.
- What it means to my owner.
- What it means to me.
- Do/can I agree with my owner about what this behavior means?

The final step for either owner or dog involves acceptance or rejection of the behavior and/or any associated meaning and emotion.

EMOTIONAL AND BEHAVIORAL STATES

In this chapter we're going to explore how the body language associated with the most common emotional and behavioral states affects our relationships with our dogs. By *behavioral state* we mean the unemotional body language as observed by the behaviorists when they study wild dogs, and by *emotional state* we mean how we think our dogs feel about that behavior or body language as well as how we feel about it ourselves. Throughout the chapter we'll compare the highly subjective emotional states (such as the body language we associate with guilt, spite, anger, and love) with the more objective and less emotional behaviors ethologists associate with the dominant, submissive, fear, aggression, or isolation behavioral states.

Before we can proceed to in-depth examinations of individual body-language signals and emotions, we need to lay a firm foundation that will permit us to relate what behaviorists consider normal body-language expression in the dog to what's typical for owner/dog interactions in general and to what's typical for you and your dog in particular. Why do we care what meaning the behaviorists attach to our dogs' behavior? Usually we don't, provided our own interpretations are acceptable, and preferably beneficial, to the relationship between us and our dogs. However, if we find our dogs' behavior unacceptable or, worse, detrimental, understanding the wild-dog roots of that behavior can help us cope with or even alter it.

For example, as we shall see in the chapters ahead, if we define the *cause* of Omicron's house soiling as meanness or spitefulness, we can correct the problem only by changing the dog's personality; in other words, we must enroll Omicron in canine therapy. On the other hand, if we recognize the behavioral functions of urination and defecation in wild dogs, we suddenly discover all sorts of solutions to the problem. By using the behaviorists' objective view instead of our emotional one, a bad dog becomes a basically good one doing something natural that, for whatever reasons, we can't accept as appropriate in our environment. Such an approach releases owner and dog to work together to change either the dog's behavior or the owner's emotional interpretation of it.

Because you're going to be encountering in the following pages many of the different forms of body language you and your dog use to relate, you might want to keep a notebook and pencil handy. By the end of the chapter your notes will give you a fairly complete list of the many body-language and emotion linkups you and your dog have already made.

GUILTY UNTIL PROVEN INNOCENT

Most owners insist that they know their dogs' guilty expressions when they see them and find it hard to believe that the behaviorists don't even acknowledge such a condition, preferring to define guilt as a peculiarly human emotion. However, no scientific explanation will ever convince the dog owner who lovingly shares pizza, designer sheets, and other strictly human accoutrements with her dog.

Whether or not our pets suffer guilt matters less than

whether or not we believe they do. For example, if Audrey Manning comes home from work and finds her miniature schnauzer, Percy, staring at her from his bed, unwilling to move and experiencing apparently uncontrollable periodic waves of trembling, Audrey might immediately respond, "Oh, Percy, what did you do? Were you a bad dog?" Audrey's subsequent search of the entire house might reveal the broken lamp, overturned wastebasket, or a telltale puddle or pile. On the other hand, Audrey's neighbor, Connie Gottshalk, associates those same signs in her cairn, Cosmo, with fear. However, if Cosmo refuses to look at her, she takes *that* as a sign of guilt and makes her own search through her home for the incriminating evidence. Meanwhile Connie's nephew thinks both women are crazy; he knows when his shepherd's done something wrong because the dog invariably hides under the bed.

From these examples we can see that there are probably as many different forms of guilt-signifying body languages as there are dogs and owners. Rather then judging them, simply note the signs you associate with guilt in your own pet.

What about the canine point of view? First, we must consider the body language we use to convey our own feelings of guilt to our dogs. It takes little imagination to recognize, again, that there are as many different signs as there are owners. When Donna Blevins feels guilty about not taking Pokey to the beach, she hugs the bulldog fiercely to her chest and repeatedly confesses how sorry she feels. Dave Potter converts his guilt to dog biscuits; the guiltier Don feels about smacking Argos or leaving him home alone all day, the more biscuits he lavishes on the weimaraner. Julie Hyman uses ice cream; Ben Davidos uses steak; Patty Yardly lets Choosie sleep on the bed.

What do you use? Treats? A special privilege? Long walks? Extra affection? Again, don't judge your behavior; simply note what body language you use to express this emotion to your pet.

BOREDOM, FRUSTRATION, AND ISOLATION

It makes sense to pair the two emotional states of boredom and frustration because most owners view the former as a more subtle, less severe form of the latter. While behaviorists almost never attribute any body language to these emotional states, they've written volumes on frustration as a *behavioral* state. Because we owners most often worry about dogs feeling bored or frustrated when left alone, we can gain a good deal of insight from the studies behaviorists have made of isolated pack animals.

Although pack animals by nature, wild dogs can and do function singularly. In fact, at times individual wild dogs function quite stably and contentedly on their own. Therefore, when we consider frustration that results from isolation, we're considering only the detrimental or negative effects that occur when one individual is separated from the pack—a lost pup, an adult trapped in a deep pit. In this sort of abnormal and threatening situation the animal may howl, whine, dig, urinate, and/or defecate to mark its territory. Howling alerts other pack members to the separated animal's location; digging serves as a means of burrowing under any restraining barriers or of constructing a shelter; territorial marking may ward off predators as well as attract the pack. All in all these behaviors comprise a set of body-language expressions we'd hardly consider negative or inappropriate, but rather practical and even quite clever responses to an abnormal situation.

Behaviorists who study *domestic* and particularly problem dog behavior agree that marking, vocalization, digging, and soiling commonly characterize the isolated dog, but they define such behaviors as negative because they can't separate the dog's behavior from the human environments. In fact, data compiled by professionals from interviews with thousands upon thousands of owners have been overwhelming—approximately 90 percent of the behavior owners consider negative occurs when their pets are isolated or alone. Though we can share the behaviorists' views of isolation behavior as beneficial and appropriate for the wild dog, most of us find this same behavior intolerable in the family pet. This leaves us with two blatantly contradictory views of identical body language. One orientation categorizes the behaviors as instinctive, unemotional, practical, and beneficial to the animal's well-being, while the other defines them as detrimental and contrary to the well-being of the pet.

Jocko may tear his footpads digging out of his wire enclosure, or worse, once he gets out, he may fall victim to a speeding car. At the very least, the owner will almost assuredly be furious with the dog when he sees the ruined pen.

But what's an owner to do? The best solution involves establishing a balance between objective and subjective realities that works best for an individual and his or her dog. If the Cormiers live on a remote hundred-acre farm, they may think nothing of their Newfoundland's howling in their absence. On the other hand, if they live in a crowded condominium complex with angry neighbors threatening to shoot their beloved Nemo, they may think of little else. Under the latter circumstances, it would probably be beneficial for all concerned to change the dog's body language, aligning it more closely with the prevailing human belief.

Do dogs exhibit any signs of boredom and frustration when they're not alone? This question returns us to subjective ground again. However, we can use the dog's behavior when isolated or alone as a clue to what types of body language we might expect at other times. For example, Marguerite Lewis's golden retriever, Aardvark, licks himself raw when left alone. When Marguerite sees him start lapping at himself when they're together, she knows it's time for a mind and body workout (physically and mentally challenging exercise) and then a swim to soothe his ticklish skin. In general this method of determining the frustration-related body language only works when the dog manifests it in some form of self-mutilation—licking, chewing, or scratching. More often than not, owners who catch their dogs displaying boredom- or frustration-related environmentally destructive behaviors quickly and soundly discipline them. Although this preserves the dining-room-table legs, it also deprives the dog of a release from the frustration it experiences. Such superficial discipline may also convert a rug chewer to a foot sucker and a hole digger to a dog who's forever digging at his ears.

Although we recognize frustration as a stronger emotion than boredom, and although frustration does cause more obvious medical and behavioral problems, boredom can also affect the quality of the relationship between human and dog. Unchecked boredom can lead to frustration; but we shouldn't automatically assume that all dogs are bored when they're alone or in an environment we humans consider boring. Trudy Seamus hates the fact that the Coughils leave their dog, Rommel, home alone most of the time. "That poor dog's bored to death," laments Trudy. "All he does is sleep when they're gone." Meanwhile Art Coughil confides to a friend that although the family was initially concerned that the dog could ever be happy in a household where everyone either worked or went to school all day. "It doesn't seem to bother Rommel. He sleeps contentedly when we're gone and then goes like a house a-fire when we're home. We can't imagine what life would be without him, and he's certainly devoted to us!"

Is it possible Trudy and Art are talking about the same dog? It certainly is. Trudy has such strong feelings about the detrimental effects of isolation that she gave her own dog away when she went back to work rather than subject it to what she was sure would be hours of loneliness and boredom. Consequently nothing the Coughils says or do will convince her that Rommel enjoys his life.

Make a list of those signs you associate with boredom and frustration in your dog. Then jot down the signs you display to your pet when you feel bored or frustrated. Are you a screamer or a sulker? Do you throw things? Does your frustration make you feel guilty? Again, don't pass judgment on your own or your dog's body language; merely note the body language you each use to convey your feelings.

AGGRESSION, ANGER, AND COURAGE

Let's put the cart before the horse for a change. Before reading on, jot down the signs you associate with aggression, anger, and courage in yourself and in your pet.

Most owners already have quite definite ideas about the types of behaviors that denote these emotions. If Paul Siegel accidentally steps on his sleeping Lhasa's tail, Paul interprets the dog's snarl and attempt to bite Paul's ankle as a typically angry response. If the Lhasa wanders outdoors to relieve itself and suddenly leaps the fence and pins the neighbor's Chihuahua to the ground, Paul's neighbor considers the behavior aggressive. And finally, when the Lhasa lunges at a huge Irish wolfhound that crosses its path during its morning walk, Paul congratulates his pet for its courage.

Upon closer scrutiny, however, these examples don't define an attendant emotion as much as they do the circumstances under which the behaviors occurred. Whenever a dog responds aggressively but does so within a context we feel is justified, we say the animal is displaying anger. If we perceive the response as protective, we say the dog is courageous. If we can't justify it, we call the dog "just plain mean."

The problem, of course, stems from some highly subjective and often subtle distinctions that may very well lie far beyond a dog's comprehension. In one of poetry's most famous lines, Gertrude Stein noted that "a rose is a rose is a rose." When it comes to aggression, dogs espouse a similarly straightforward view: Aggression is aggression is aggression. Their world doesn't acknowledge our value judgments, such as good or bad, appropriate or inappropriate aggression—only that which succeeds and that which doesn't. In other words, aggression falls into the behavioral rather than the emotional category. Furthermore, the signs we often link to this state—growling, snarling, ears and tail erect, rigid posture, hackles raised—look just like those the behaviorists link with normal dominant behavior.

When does dominant behavior become aggression? Dogs use dominant and submissive body language to keep peace within the pack or between two individuals. If erect ears and tail and growling achieve that purpose, the animal has successfully expressed dominance. If the display initiates a fight, then the identical body language now signals aggression.

Interestingly, the more dominant one animal behaves toward another, the less likely it will find itself in a combat situation. When Lou Rutherford gloats over his black Lab's almost daily skirmishes with his neighbor's mastiff as proof of Merlin's superiority, Lou doesn't realize that the altercations actually indicate the opposite. Merlin isn't the dominant dog at all. Because the dogs perceive themselves as equal, they keep fighting. The dominant-dog award actually goes to the black Lab across the street, who rarely has a mark on him, the dog Lou describes as "mellow yellow."

Owners who unknowingly encourage dominant behavior in young pups can provoke later confrontations with their own pets. If we encourage our Great Dane pup to put his front paws on our shoulders, how are we going to respond to his growling and snarling when we try to push him down eight months later? What we thought was an invitation to play was really an invitation for dominance; when we challenge that dominance, the pup responds aggressively, and we become angry.

Finally we must note that most of us use the terms *aggression* and *hostility* synonomously. *Aggression* actually refers to the willingness of one animal to respond more quickly and successfully to a particular set of circumstances than another. If those circumstances center around a territorial dispute, the body-language signals may appear hostile. In other situations, though, aggression may manifest as the ability to hunt longer, respond more quickly to game in flight, analyze scent data more efficiently, or stalk prey or raise young more patiently.

Review the list you've made of body language you associate with aggression, anger, and courage in yourself and in your pet. Are the three emotions and their signs as easy to define and separate as you thought?

FEAR

Much has been written about the body language of fear because it's one of the most easily recognized responses in all species. Furthermore, most of us recognize fear as an emotional as well as a behavioral state. For example, Margot Geoghegan is afraid of spiders. She thinks of them as being the size of dinner plates with an infinite number of furry legs, and the very thought evokes feelings of fear. When Margot actually sees a spider scurrying down the front of her blouse, her pupils dilate, she screams, jumps up and down, and pounds her chest repeatedly; these are the body-language signals she uses to convey her fear. She may have a completely different kind of response when she thinks of losing her wallet or if the brakes on her car should fail; and the body language associated with the actual occurrence of these two events may be entirely different. Yet in Margot's mind all are undeniably valid manifestations of fear.

Why does fear elicit such a powerful combination of emotion and behavioral response? If body language must serve some purpose in order for it to persist, the purpose of fear must be a powerful one. And it is. The purpose of fear is to protect the individual from harm. Furthermore, as letter carrier Joe Spivak learned, we can benefit from learning to recognize fear in others as well as in ourselves. Once Joe recognizes that Pumpkin considers him threatening, he can diminish or eliminate that fear with a little game of ball.

Behaviorists have long held that dogs express their fear in one of three ways: they freeze, they fight, or they flee. Let's take a closer look at the specific body language of each of these responses. Obviously, if the fearful dog disappears, the major signal we see is its tail vanishing into the shrubs. By running away from what it perceives as a threat to its existence, the dog conveys one of two messages: (a) that it lacks sufficient experience to know whether to respond to the threat in a dominant *or* submissive way; or (b) that it has sufficient experience to recognize that this threat will force it into an unacceptable position. In any event, the dog perceives running away to be in its best interests.

Dogs that feel they can best protect themselves by freezing or fighting offer abundant and consistent body-language cues. From our earlier discussion of dominance/aggression and submission, we know that a dominant animal will feel more confident about fighting when frightened, whereas a submissive one will tend to freeze under the same circumstances. However, the wise owner takes such generalizations with a few kernels of kibble. Because the behavior states of dominance and submission result from interaction with others, experience plays a pivotal role in determining fear reactions. Even the most dominant of dogs may roll over and pee submissively at the sight of an old man on crutches, and a shy sheltie who tucks in her tail at the sight of any other dog may dominantly herd a group of pre-schoolers across the street.

What other signals telegraph fear? A close inspection of canine behavior reveals two distinct categories:

I	II
Soft growling	Growling leading to short, loud barks
Ears flattened against head	Ears erect
Tail tucked against belly	Tail rigid, straight out, or slightly up
Stiff-legged stance	Body slightly lowered, rear legs braced
Urinary dribbling	Deliberate urinary marking
	Hackles raised

Depending on both a dog's personality and the specific fear-provoking situation, a dog may exhibit either category of behavior. For example, when Dory Humbolt stays home alone with her children and an unfamiliar car pulls into the driveway, the Humbolts' malamute, Sasha, exhibits classic category II behavior. However, when Sasha's male littermate violates her territory, she expresses typical category I body language. Were she to encounter her male littermate in his own yard or on some neutral ground, she'd drop to the ground, roll over submissively, and permit him to circle and sniff her without fear. But her protective attachment to the Humbolts makes her perceive him as a threat in her own yard, and consequently she appears more willing to defy the male-female dominance and submission roles that maintain peace at other times or in other places. When Dory's husband stays home, Sasha permits her littermate to enter the property without challenge and accepts his dominance because she views the adult male human as top dog. In such a situation she feels no need to defend the territory. As these examples show, Sasha does not so much perceive a threat to herself as to her owners' property. However, she responds exactly as she would if she considered herself in danger.

Obviously the body language of fear doesn't follow hard-and-fast rules. Fear responses vary not only from dog to dog but also from situation to situation. Review the two lists, recalling your own dog's reactions to various familiar and unfamiliar situations. What signs does your dog most often display? Is your dog more likely to freeze, fight, or flee? How do your own responses to threatening situations compare with those of your dog? Do you consider your dog too aggressive or too timid?

No discussion of the body language of fear can be complete without recognizing the clear connection between fear and aggression. Aggression is the dominant animal's common response to fear. It is the fight response in action.

How do you feel about that? Because many owners differentiate between aggression (and its more noble form, courage) and fear, they react to fear signals in two completely different ways, depending on the situation. Doing so sends inconsistent body-language messages to our pets and could cause a host of problems.

For example, when Cathy Colby spends a week alone in her parents' secluded summer home in Maine, she's pleased to have Cromwell, her English setter, at her side. Every time the dog barks or growls, assumes a rigid stance with hackles raised and ears erect in response to some strange noise, Cathy praises him for his courage. When Cromwell exhibits similar behavior in response to delivery persons, doors slamming, sirens, and other familiar sounds surrounding Cathy's mid-Manhattan apartment, her typical response is "Oh, Cromwell, stop being such a baby. Shut up and go lie down!" If she succeeds in "training" him to ignore novel sounds in Manhattan, how likely is it he'll respond in Maine? If he doesn't respond in Maine, will Cathy feel her courageous dog has become a coward?

SUBMISSION, DEVOTION, AND COWARDICE

"Kneehigh loves it when I tickle her tummy" insists Stella Mann.

"Ace can't let an evening go by without having his belly scratched," claims Peter Hood.

"This chicken hound of mine goes belly-up every time she sees the Yorkshire terrier next door!" fumes Carl Alern.

"All I have to do is look at Chicory crossly and he rolls over and pees," says Ted Lawrence in exasperation as his dog drops like a rock.

"Did you ever see such devotion?" Bonnie Howard asks her husband as they observe their bassett, Zonkers, with their youngest child. The dog's adoring gaze never leaves the little girl for an instant.

Do any of the foregoing statements strike you as the least bit unusual? Most owners have experienced some or all of these phenomena within the context of a perfectly normal relationship with a pet. However, as we so often do with expressions of dominance, we usually assign emotional definitions to these submissive body-language displays. Stella, Peter, Carl, and Ted observe the same exposed abdomen position, yet Stella and Peter view it as a sign of love and even intelligence, whereas Carl and Ted view it as evidence of their pets' cowardice. While neither Stella nor Peter would relinquish their subjective interpretation in favor of the behaviorist's unemotional one, perhaps Carl and Ted could use the behaviorists' objective view to dissipate their negative feelings about their pets' behavior.

What kinds of behaviors do you associate with submission or cowardice in your dog? What behaviors do you use routinely when you give in or submit to your dog? (I'm a fan of the heavy-sigh-clenched-teeth routine: *"Oh, all right, have it your way!"*)

"Give in to my dog? Never!" exclaims Ferdi McCann. But Ferdi forgets the times she shoves Tiberius out the door in frustration over his barking, the times she responds to his pawing by putting down her needlework and petting him, and those times she yields to his imploring gaze by giving him half of her ham sandwich or all the leftover canapes from the party.

Another interpretation of submissive behavior, devotion, is 100 percent subjective. Behaviorists acknowledge no signs of canine devotion, yet every owner can easily recognize it in his or her pet. Most often it involves a direct gaze, sometimes an attentive tilt of the head, usually while the dog is sitting or lying down. In addition, most owners at least occasionally experience a sense of awe or reverence toward their animals in general or toward some wonderful deed the pet has performed. Hunters talk of being overwhelmed by the sight of their bird dogs streaking through the fields. Some border-collie owners get choked up every time they see a good dog work. Returning college students feel awe when the family dog instantly throws itself into their arms. Some of us just can't get over how happy our dogs always are to see us.

Does your pet worship you? How can you tell? Are you in awe of your pet? What does your dog do to elicit your feelings of reverence? How do you signal your own reverence for your dog?

PATIENCE

Many would argue that patience is not an emotion, let alone an emotion exhibited by dogs. However, we're going to consider it because its presence or absence in a relationship acutely affects the quality of any dog/human interaction. When Debbie Arno leaves home at 7:00 A.M., seldom to return before 5:30 P.M., she needs a patient dog; and when Dasher, the dachshund, topples the Christmas tree trying to get at the candy canes dangling from its branches, he needs a patient owner.

The most common sign of patience in dog and owner is acceptance, but how do you detect such a passive trait? Unfortunately it seems we only recognize patience by its absence. Dick Burgess berates himself for losing his patience and smacking his pup when it refuses to obey a command. When her Boston terrier, Seagram spits out his medication for the eighth time, Paula Petchak grabs him and shakes him like a mop. Cookie whirls and snarls at a visiting child who has been tugging on her ears. Chimmi tires of Tom's repeated yanks to remove burrs from her silky coat and begins growling softly. So perhaps we can define patience as lack of emotion or, better, a calm emotional response to body language that could elicit a highly charged one.

Patience provides a safe harbor, enabling owner and dog to tolerate and accept stormy events and behaviors that would otherwise cause anger and even violence. It may be the hardest emotion to elicit at will because more often than not we spend much time and energy trying to avoid those difficult situations where we could use it. Then when these situations occur, we lack the experience to cope with them any way but erratically.

If personal fears kept Paula from handling Seagram's mouth when he was a pup, trying to accomplish this task when the dog requires oral medication creates a number of negative emotions:

- She's still afraid of sticking her fingers in his mouth.
- She feels guilty that her ineptitude may prolong his illness.
- She's angry he put her in this position.

If Tom never grooms Chimmi unless her health depends on it, the eventual grooming will turn into a time-consuming and even uncomfortable process during which both dog and owner can easily give in to negative feelings.

Let's look at one more common example. Josie Carney has no patience with the way her Airedale, Jake, behaves around other dogs. She finds the show of dominance, the posturing, circling, and sniffing repulsive. Consequently she fenced in her yard and forbids Jake to leave it or mingle with other dogs. Whenever Josie takes Jake to the veterinarian for his annual inoculations and checkup, bedlam breaks loose. After the last episode, Josie vowed never to take the dog anywhere.

Because Josie had no understanding of and patience with the original behavior, she avoided it rather than either accepting or changing it. Now it will require a great deal more patience from both dog and owner to deal with what has mushroomed into a major problem.

Patience doesn't mean martyrdom. When Dave Talman takes his all-male Welsh terrier for a walk, the dog lunges at every other dog and lifts his leg on every tree, fire hydrant, and (Dave's convinced) blade of grass. Dave sighs heavily and cringes noticeably every time his neighbors observe the dog's loutish behavior. "What else can I do? After all, he's just being a dog," he says with the air of a condemned man. Is Dave truly exhibiting patience with his dog, or has he let himself be controlled by him? Once again, an emotion has become inextricably entangled in the owner/pet relationship. Does being patient and accepting mean we let our pets control us? When your dog exhibits patience with your foibles and inconsistencies, do you view that as part of its rightful subordinate position? Review situations in which you or your dog displayed patience, noting the behaviors displayed and any results. Also determine whether you consider yourself more or less patient than your dog.

SPITE

"CoCo chews up my hat just to spite me!"

"Snippet won't eat because she wants to punish me for not getting her favorite food."

"Romulus peed on the rug to get even with me for going out with Tod."

"Hercules whines to get back at me for having the stereo turned up too loud."

Isn't it funny that we so frequently attribute spite to others but not to ourselves? And although we can easily detect signs of even so "negative" an emotion as anger in ourselves, few of us like to think of ourselves as spiteful—even though we couldn't spot it in others if we didn't know it well enough in ourselves. Spite is probably the most complained-about emotion owners attribute to their pets.

Although owners perceive a vast collection of different behaviors as evidence of their dogs' spitefulness, behaviorists again refuse to accept that such a state exists in dogs. Members of wild-dog packs simply don't show evidence of spiteful behavior toward each other, probably because, as we noted, wild animals only perpetuate those behaviors that serve some beneficial purpose. Because we perceive spite as someone doing something against us purely for the devilishness of it, it could hardly fulfill any real survival purpose. In fact, wild dogs engaging in such displays while others were hunting, mating, or rearing young would most likely be eliminated.

Regardless, most owners do feel that their dogs can and do display spiteful behavior toward humans, perhaps because the canine companions have learned this form of communication from their owners. One fine example was Suki, a five-year-old castrated male miniature poodle deeply devoted to his owner. In general, Suki was a perfect companion to his owner, and the two led a mutually rewarding life together. Still, Suki found one of his master's habits intolerable: If his owner left home for more than two hours, Suki would systematically pull record albums off the shelf, drop the records on the rug, and chew the jackets. Astonishingly, the dog never touched anything else, even though he had access to more tantalizing and more conveniently located objects. Destructive isolation behavior? Maybe, but Suki's owner was an ardent record collector and prized his collection above everything but the dog. If someone (or some dog) wanted to get even with him, the records made an ideal target.

Spite, like guilt, has a way of becoming a decoy emotion. If we assume it causes a certain behavior, where do we go from there? Canine psychotherapy? If Choosie pees on the rug out of spite, then we can only resolve or change this behavior by despiting her. Furthermore, a person who believes he or she is the target of spiteful behavior invariably finds it impossible to respond objectively. More often than not, if we sense spite in our dogs, we respond in one of two ways:

- If we agree that the spite is justifiable, we feel guilty.
- If we feel that it isn't justifiable, we get angry and want to retaliate, to get even with them for wanting to get even with us.

Grit your teeth and list all the behaviors you consider signs of spite in your pet. Then list those qualities in your own behavior your dog might interpret as spiteful.

HAPPINESS AND LOVE

Pick up any book on animal behavior and look for the word *love* in the index. I've never seen it beyond the dedication page, where the behaviorist thanks his or her spouse or parents for their support during the writing of the book. About the closest they come to a description of the behavioral state of love is when they discuss mating; but since few of us view our pets as sexual partners, this interpretation hardly seems applicable. However, pick up any book in which an owner describes interaction with a pet and you'll find the word *love* on every other page.

No other emotion serves a more critical role in the relationship between owner and pet. Even if every behaviorist in the world were to line up and shout, "Dogs can't feel or express love!" no sensitive owner would agree or stop acting as if dogs can love others. In fact, I know many behaviorists who compile data totally and deliberately devoid of any emotional overtones and then go home to interact with their pets in what can only be described as a loving way, one that bestows happiness on both owner and dog.

If no scientifically recognized signs of love and happiness exist, how do we know our pets love us? We could say, "We just know," but most of us need something more. Some common behaviors or body language owners associate with love and happiness in our dogs include the following:

- Marla Iglesias feels her Shih Tzu, Kung Foo, loves her and is happy if he instantly gobbles up all the food and treats she offers him.
- The Chapmans believe their poodle loves them because he sleeps on their bed.
- Carrie Sinclair knows Max loves her because he always obeys her commands.
- Hal Dinsmore insists his Brittany loves him because she attacks anyone who comes into his yard.
- Jeff Copeland feels his dog loves him because the sheltie mix refuses to obey anyone else.
- Marge Warren knows Corky loves her because the corgi always leaps and dances frantically when she comes home.
- Jon Hart thinks Exador loves him because the black Lab destroys the house when they're apart.

- Jane Conant knows her dog loves her because the golden wouldn't dream of doing anything destructive regardless how long she's alone.
- The Rutherford dogs wag their entire bodies and leap ecstatically to show their love.
- Chester Ball's old bloodhound utters a special happy sigh.

This list could continue indefinitely, and it would reflect many of the many-splendored signals people cite as expressions of love between humans.

In addition to the infinitely variable range of body language we consider expressions of our dogs' love, we owners have armed ourselves with an equally extensive repertoire of body language to communicate our own love and happiness to them:

- Dolly DeLeo hugs and kisses her poodle the same way she does her three-year-old son.
- Alan Bates gives Chaunny a food treat the first thing each morning and last thing at night.
- The Hapgoods share their love with their Belgian shepherd by taking him for rides in the car.
- Bob Dufour lets Teddi sleep by his bed.
- Charlene Roylston feeds Mimi sirloin steak.
- The Armitages have an intricate set of whistles and claps they use to let their Yorkie know he's made them very happy.

Given the infinite variety of signs associated with the expression of love and happiness, every owner and every dog is free to develop any that serve their relationship. What body language do you consider signs of happiness or that your dog loves you? What ways do you use to show your love and happiness to your dog?

SADNESS, SORROW, AND DEPRESSION

Of all the qualities owners attribute to their dogs, most people cherish most their dogs' uncanny ability to recognize when they're depressed. Every dog owner could cite a collection of anecdotes about how their pets immediately knew they'd lost the contract, broken up with a lover, had a fight with a best friend, or didn't get the anticipated raise. Even more than love, sorrow seems to exude from us humans in a form most dogs seem capable of sensing, perhaps because we communicate such obvious cues to this emotional state. We frown, we sigh deeply, we hang our heads and cry. Scientists have discovered that the chemical composition of tears of sadness differs distinctly from that of tears resulting from irritation; maybe our canine companions with their incredible olfactory apparatus can even *smell* our sorrow.

For all the dog's seeming ability to detect sadness in us, we reciprocate poorly. It's rather ironic that descriptions of many illnesses in veterinary texts list depression as a valid symptom, yet recognizing this often critical sign requires finely tuned intuition. Remember the comedy routine in which a sad-sack basset hound sits with a fixed mournful expression while the comedian proclaims that the dog is actually deliriously happy? Yet well-bonded owners, veterinarians, handlers, and trainers *can* easily tell the difference between a physical appearance that mimics sadness or depression and the real thing. Nor are they fooled by outward signs of happiness that mask a depressed or unhappy animal. Sad or depressed animals will ignore their food and take very little interest in their surroundings. If you urge them outdoors, they'll go, but you can tell they really don't care whether they do or not. Favorite toys and/or rituals trigger little or no response. They select that spot in which they feel most secure and sprawl there for hours. So while the telltale signs in the animal's body language may be limited or subtle, the overall impression they give quite plainly communicates depression.

Therefore, the recognition of this emotion requires a great deal of sensitivity and confidence in our relationship with our pets. For example, Joe Overmeir doesn't hesitate to take his English bulldog to the vet based on his intuitive awareness that "Winston's not his happy old self." Rarely, if ever, does Joe err; the vet usually discovers an infected ear or irritated skin fold. Furthermore, Joe feels he can differentiate between medical and behavioral unhappiness. If he suspects the latter (Winston misses his collie playmate whose family moved last week), he takes his dog for three daily walks instead of their usual two and makes a special effort to interact with his dog rather than rushing him to the vet.

Although well-bonded owners can detect subtle changes in their dogs' appearance or body language that signal sorrow or depression, others create problems for themselves and their pets when they erroneously relate certain signals to these emotions. For example, Kathy Crossland is convinced her sheltie mix grieves constantly whenever she and the dog are separated. Kathy cites her dog's "mournful brown eyes" as evidence of the dog's emotional condition. However, the dog displays no behavior that substantiates this view—she eats well, enjoys romping with her owner, and displays no negative medical or behavioral symptoms. What the owner has zeroed in on as a sign of her dog's sorrow—those big brown eyes—are merely the typically expressive eyes of a normal sheltie.

How do you commonly express sorrow? How does your dog respond to your sad moods? How does your pet convey its sorrow or unhappiness to you? How do you respond to its body language and emotion?

IT'S ALL RELATIVE

In this brief overview of the most common emotions that season our relationships with our pets, we've uncovered very few solid facts. We did see how we can take the behaviorist's broad categories of dominant and submissive body language and apply them to various emotions we recognize in our dogs. We also learned that emotional responses and body language are highly dependent on individual differences and situations. We may be able to say Percival responds joyfully to a whistle, but few of us would be willing to guarantee that he *always* will. We know he won't respond if our imposing next-door neighbor's in the room.

Does this apparent lack of solid scientific foundation mean

that body language and emotions are so highly subjective as to be meaningless? Indeed not. It does mean, however, that we must learn to read our dogs as individuals and pay attention to the varied and surprising forms body language and emotion may take in both human and canine. By now you should have a good idea what emotions and body language you and your dog currently use to communicate. In the next chapter we'll see how we can use that knowledge to solve problems and enhance our relationships with our pets.

3

USING BODY LANGUAGE AND EMOTION TO SOLVE PROBLEMS

ONE PARTICULAR MONDAY everything went wrong for Mary-Ellen Howell. The dishwasher exploded, destroying her best china, her arms and legs broke out in itchy red welts—her reward for clearing poison ivy out of the garden Sunday—and her seven-year-old twins awakened that morning with stomach pains. How could Mary-Ellen possibly get the house clean and the shopping done before her husband's boss and his wife arrived for dinner? And what about her yet unfinished half of the project her business partner planned to pick up that afternoon?

"Mary-Ellen, Magoo and I have come to help," called out a cheery voice from the back porch. A forty-five-pound ball of white fluff hurled itself at Mary-Ellen's bare legs.

"Leigh, can't you teach that damn dog to stay down?" cried Mary-Ellen, angrily swatting at the dog.

Leigh Heisman yanked the bewildered dog away from her sister. "What's the matter with you? You used to enjoy Magoo's greetings."

Mary-Ellen could hear the hurt in her sister's voice. Magoo looked from one woman to the other as if to say, "People! You can love 'em, but you sure can't understand them!"

* * *

In a hectic household like Mary-Ellen's, an exuberant dog can be the straw that breaks the camel's back, and yesterday's approved behavior can become tomorrow's canine sin. Let's consider the not uncommon or very complex interaction between Mary-Ellen, her sister, and Magoo. Who or what is the real problem? Mary-Ellen's Monday morning blues? Magoo's rowdiness? Or Leigh's failure to train and discipline her pet? In a famous Japanese film, *Rashomon,* a crime occurs, and the three people involved all remember it quite differently, depending on their own unique perspectives. In the same way a lot of simple canine body language may mean many things to many people (and dogs). To Leigh, Magoo's jumping tells her he's happy and healthy; to Mary-Ellen it can be a source of pleasure in some instances, a source of irritation in others; and to a deliveryman who gets toppled down a flight of stairs, it can be cause for a lawsuit. To Magoo, these conflicting human reactions are simply a source of confusion. To make sure you and your dog create and perpetuate body language and attendant emotions that enhance rather than damage your relationship, let's learn how to dissect a human-canine body-language and emotion problem.

RECOGNIZING THE PROBLEM

Although most of us recognize a problem within the context of our own personal definitions of good and bad behavior, we often neglect to explore it beyond this highly subjective point of view. To be sure, we should heed our intuitive inner awareness, but we must also accept that other people, even those closest to us, may not see things the same way we do. For example, while Mary-Ellen, Leigh, and Magoo all feel that something is wrong, their definitions of the problem vary dramatically:

- Mary-Ellen believes Magoo is wrong for jumping on her and that Leigh's wrong for permitting the dog to do so.
- Leigh feels it's wrong for Mary-Ellen to hit Magoo and accuse her of condoning Magoo's bad behavior.

- Magoo senses there's something wrong because both women are ignoring him.

Every problem arising between human and dog can originate with the human, the dog, or both. Unfortunately, most humans begin with the assumption that "I'm okay, the dog's wrong," or "The dog's okay, I'm wrong." Such simplistic views not only hamper the development of a relationship, they also severely limit possible solutions when problems arise.

Whenever we believe a canine problem exists, we must first determine whether it's a physical or a behavioral problem. Obviously Mary-Ellen believes Magoo has a behavioral problem. However, if Magoo were a geriatric fourteen-year-old rather than an exuberant five-year-old, she might easily wonder if the aging dog had begun to lose his sight.

What if we find a particular behavior acceptable but others disagree? In such instances we usually dismiss them as intolerant dog-haters. Mary-Ellen questions her sister's relationship with Magoo, and Leigh finds such reproach intolerable. By the same token, Leigh feels Mary-Ellen could improve her relationship with Magoo, an opinion Mary-Ellen doesn't necessarily share.

Emotions frequently cloud an objective assessment of any problem. Although we may fault Mary-Ellen's emotional and body-language response to Magoo's jumping, she can't possibly resolve it if she only sees the dog once a week. Any consistent changes will depend on Leigh dispassionately accepting the behavior as a problem; if she refuses to believe a problem exists, the behavior will persist.

DEFINING THE PROBLEM

Once we recognize that we or our dogs have a problem, we need to break it down into its body-language and emotional components. Let's analyze what commonly occurs when an owner tries to teach a pup to walk on a leash. One sunny June morning when Magoo was eight weeks old, Leigh put a bright new red collar and matching leash on her Samoyed and carried him to the park. Positioning herself in front of the pup, she looked him squarely in the eye and said, "Come," while tugging on the leash. Magoo averted his gaze, splayed out his front legs, and dug in with his rear feet, resisting all movement.

"Magoo, I said *come!*" Irritation hardened Leigh's voice as she pulled more firmly on the leash. Magoo promptly flattened himself and was dragged along, his rear paws plowing up tiny furrows in the soft ground.

"Magoo, you listen to me! I SAID COME!!" The sound of Leigh's voice and her rough jerks on the leash overwhelmed Magoo. He rolled on his back and urinated submissively. Leigh, torn between anger at her pup and frustration and guilt regarding her own behavior, broke down and cried.

We can separate problems like this into four basic parts:

- The dog's body language.
- What the owner or others believe the body language expresses.
- The human emotional response to the dog's behavior.
- The human body language used to express that emotion.

Such a breakdown helps us separate even the most complex interactions into something manageable. All owners experience exchanges similar to the one between Leigh and Magoo, hodgepodges of actions and reactions that can easily distract us from what's really happening. Only by objectively defining each component can Leigh eliminate her unproductive emotional judgments and get to the heart of the problem. Leigh knows something is wrong, but her anger and tears prevent her from getting to the root of the problem. Magoo may also know something is wrong and that it has something to do with Leigh, but at this age he responds to his instincts more than to his owner.

What does Leigh believe Magoo communicates through his behavior? "At first I thought he might be scared, but that's silly because there's no reason for him to be afraid of a leash and surely he's not afraid of me—he knows I love him. So I decided he was just being stubborn and spiteful, trying to test me to see if he could get his own way. But when he rolled over and peed, I thought maybe he really was scared." Leigh believes Magoo's body language reflects his fear, stubbornness, and spite.

However, what body-language message did Magoo really express? First the body language:

- He averted his gaze.
- He became rigid.
- He flattened himself against the ground and refused to move.
- He urinated submissively.

When these body-language signals are separated from the rest of the interaction, we quickly see that most of them reflect submission and a submissive response to fear, the freeze response.

How did Leigh respond to Magoo's emotional and body-language displays? "Initially I thought it was normal, so I was patient; but when he kept it up, I got frustrated and then mad at him. When he rolled over, I felt guilty that I might have been expecting too much from the poor little guy. That made me feel really sorry for yelling at him." In response to Magoo's fear, spitefulness, and stubbornness, Leigh experienced patience, anger, frustration, guilt, and sorrow.

Finally, what body language did Leigh exhibit in response to her pup's behavior and her own emotions? She yanked the leash harder and longer, her voice became louder and more threatening; finally she gave up and cried. Although Leigh experienced a wide range of emotions, the body language she displayed communicated dominance to her dog.

By isolating a problem's components we can often see where body language exchanges between human and dog, like those between dog and dog, often reflect little more than expressions of dominance and submission. Furthermore, we can see that the interspersion of various emotions among the dominant and submissive physical cues can often lead to incorrect and even harmful interpretations. In essence, Magoo kept responding to Leigh's dominant body language submissively, instinctively trying to keep the peace. However, instead of accepting his behavior (and leaving him alone as another dog would), Leigh flashed more and more dominant signals.

Imagine a thief who demands, "Give me all your money!" And after you hand over your wallet, he screams, "GIVE ME ALL YOUR MONEY!!" Having already done so, you would probably fear for your life at the hands of an irrational madman. Your life depends on doing *something,* but fear freezes you in your tracks: Should you try to dominate him? Should you try to run for it? Or should you try to find some way to convince him that you would do anything *within your ability* to save your life?

When Leigh commands Magoo and tugs on the leash, putting pressure on his neck area (a dominant location), he interprets that as "Submit!" When he reacts with the averted gaze and rigid freeze response of submission, Leigh transmits body language he interprets as "Submit more!" We can easily imagine what a total fiasco occurs when owners become so frustrated they beat the dog under these circumstances. Christian theology asks the rhetorical question, "If your child asks you for bread, who among you would give him a stone?" Owners who respond to submissive displays with violence should remember these words.

POINT OF VIEW

Having defined the problem in terms of both human and canine body language and emotion, we must further explore the multifaceted nature of point of view: Who's defining the problem and what relationship does that person have with the dog? In the leash problem, who defined the emotion and body language being expressed? "Obviously *I* did," says Leigh. That's right, but most of us often overlook the implications of our role in this process. Suppose instead of saying that Leigh defines Magoo's and her own body-language and emotional responses, we claim that she imagined them.

"I didn't!" Leigh hotly retorts. "I know what Magoo did. I

saw him. I might not know why *he* did it, but I do know why *I* did what I did."

Leigh's remarks indicate a healthy attitude, but other owners faced with similar challenges often take something so highly variable and subjective as the interpretation of an interaction with their pets and try to foist it off on some outside source:

> *"They* say only mean dogs do that."
> *"They* say people who act like that shouldn't own big dogs."
> *"They* say little dogs shouldn't be allowed around little children."

If we could ever track down this omniscient "they" perhaps "they" could magically solve all our problems with our dogs. Actually, exactly the opposite would probably happen. Whenever we accept a definition of a problem remote from our own experience and beliefs, that definition has little lasting significance for us; and the less significant we find it, the less likely we are to implement any consistent changes long enough to solve the problem. For example, suppose you watch an expert dog trainer on television who always demonstrates his training techniques using a beribboned five-pound Yorkshire terrier, but you own a hundred-pound bloodhound who's never been inside your house. When your dog develops problems, whose advice are you most likely to heed, the famous trainer or the fellow down the road who's raised hounds for years?

Consider another aspect of point of view: *Whose* problem is it? We already noted that it can be a dog problem, an owner problem, or another person's problem; it can also be an environmental problem. Let's see how Leigh evaluates the origin of her problem with Magoo:

"It's a dog problem because Magoo's too shy, but it's also my problem because I wasn't patient enough with him and didn't understand what he was telling me. It was an environmental problem, too, because I took him to the park for our training session, but he'd never been there before, and a gang of kids on bikes may have confused and alarmed him even more."

By recognizing how different individuals and factors can contribute to completely different interpretations of what's going on, Leigh's well on her way to trimming away extraneous and negative elements from her interaction with Magoo.

"I think I'll accustom Magoo to the leash at home before I take him into more challenging environments," says a confident Leigh. "My sister might have been able to train her dog in the park, but Magoo's a different dog and has different needs." She now recognizes the importance of creating an environment tailored to Magoo in order to simplify rather than complicate their exchange of body language and emotion.

Fortunately, Leigh and Magoo didn't have to deal with someone else's point of view in this particular instance, but they will in many others. Any time we set about changing behavior because someone else thinks we should, we start out at a tremendous disadvantage. Whether a problem is defined by a famous trainer, author, veterinarian, relative, or friend, if any aspect of the definition strikes us as untrue or not right for us and/or our pets, our chances of effecting any permanent changes plummet. When Leigh's sister lashed out at behavior she'd accepted in Magoo since puppyhood, Leigh decided to let the matter drop until the more immediate crises in Mary-Ellen's life were resolved. However, she called her sister several days later to learn whether Magoo's actions had truly offended Mary-Ellen. Although Leigh herself enjoyed the Samoyed's exuberant leaps, she didn't want the dog's behavior to jeopardize her relationship with Mary-Ellen and her family. As it turned out, Mary-Ellen *did* find the behavior troublesome but had been unwilling to mention it lest she offend Leigh. At first Leigh was shocked and hurt by this revelation, and had she initiated any program to stop the jumping "because that's what Mary-Ellen wants," her relationship with both her dog and her sister would have suffered. However, she gave herself several days to mull over her sister's comments, decided they were valid, and began a training program because *she* believed it was the best course for her and Magoo.

If you find yourself in a similar situation, honesty is the *only* policy. If the professional trainer interprets your dog's behavior in ways you can't accept or advocates methods of training you can't adopt, say so. If your veterinarian reads things into Sniffy's behavior you believe aren't there, feel free to voice your opinion. If your sister or nephew encourages your dog to exhibit what you consider inappropriate body language, ask them to stop.

Such advice may be easier to give than to follow. However, it's surprising how often owner/dog relationships deteriorate because owners struggle to make their dogs conform to someone else's definition of a healthy or well-behaved dog, a definition that they themselves do not accept. Unless we feel comfortable with others' interpretations of our dogs' body language, any changes that need to be made, and the ways in which those changes should be accomplished, we can't achieve lasting results.

However, because the interpretation of body language is so subjective, we can have many different interpretations for the same body language; so how can we be sure our interpretation is the best one for us and our dogs? Let's look at an example: Once I tried for months to get a client to put her dog on a diet. I predicted an ominous future filled with premature arthritis, cardiovascular collapse, and digestive ills, all to no avail. Then one day I saw my client briskly walking her obviously slimmer and more vigorous pet.

"My, but Rollo looks trim today," I observed, modestly preparing to accept her thanks for the dire medical predictions that had induced her to follow my advice.

"My neighbor called him a fat pig!" she said indignantly. "That did it! I cut out all the treats and began exercising him more every day. Don't he look grand!"

Indeed he did. Rollo had lost weight and gained several extra years of active life. For my part, I'd learned a lesson in humility and the value of an owner's support if any change is to be accomplished. Although Rollo's owner never doubted the truth of my advice, she believed Rollo was perfectly healthy and happy, albeit maybe a bit pudgy. When her neighbor called him a fat pig, the message Rollo's extra pounds conveyed to others carried a far more negative charge than my dire predictions. Unlike my rather obscure medical interpretations, which might only be shared by those within the medical community, the neighbor's could easily be shared by everyone on the block. The fact that her friends believed Rollo's extra pounds resulted from his piggish nature finally convinced the dog's owner to alter her own and her dog's behavior.

In this example the canine body language in its purest sense communicates overweight; Rollo should weigh twenty-five pounds, according to all medical data and breed standards, but he tips the scales at thirty-two. This body language is interpreted three different ways by three different people:

ROLLO'S OWNER: "He looks perfectly normal to me."
VETERINARIAN: "He's obese and a candidate for all sorts of medical problems."
NEIGHBOR: "He's a fat pig!"

Looking at these evaluations *objectively,* none is more or less valid than any other. However, each is considered the *most* valid by its originator and anyone who supports his or her views.

What happens when the owner's interpretation is in conflict with those of others? Often it depends on who those others are and what we think of them (first) and their opinions (second). For example, although we may all want the most expert advice, sometimes we find it difficult to associate such advice with our own view of the problem. If we respect (or fear) the expert sufficiently to do what he or she suggests without understanding why, we may be able to accomplish the desired change. However, more often than not we'll never be totally committed to the change. Had I been able to convince Rollo's owner to cut back on his food even though he looked and acted perfectly fine to her, chances are she would have done so for *me* rather than for herself on her dog. That being the case, as soon as she stopped seeing me at regular intervals, her reason for keeping Rollo slim would vanish and he would blimp out again. Although getting the weight off to stop her neighbor's negative comments about her beloved Rollo shouldn't, in theory at least, be any more successful, the fact that she respects her neighbors and sees them every day reinforces her choice to put Rollo on a diet and make it work for her.

Suppose, however, Rollo's owner finds both her neighbor's and my interpretation of her pet's body language unacceptable? If these other evaluations bother her, it will be for one of two reasons:

- She believes these other interpretations are incorrect.
- She believes they are correct but doesn't want to accept it.

In general we want people to like us and our pets—or at least not dislike us. Therefore, when others make what we consider erroneous comments about our dogs, it bothers us. Although we may try to ignore it at first, if we truly enjoy our pets and want to share that joy with others, we may choose to attempt to alter the other's interpretation. For example, many people have blanket prejudices regarding German shepherd dogs and Doberman pinschers: any body language these dogs express is viewed as threatening. Because this is such a prevalent view, many of the people who fancy these breeds go out of their way to produce gentle, well-behaved pets. In this case, the incorrect view of others spurs owners to do positive things they might not otherwise do. There is also that minority who, rather than ignoring or attempting to change another's incorrect interpretation of their dog's body language, will seek to make it true. "Oh, so you think all Dobermans are mean. If that's the way you want it—'Sic him, King!'" *Why* owners do this falls more into the realm of the psychologist or psychiatrist then the trainer, handler, veterinarian, or behaviorist. However, the result—a dog exhibiting the very behavior the owner wants to believe is nonexistent—makes it a canine as well as a human problem.

Others' interpretations of our dog's body language can bother us even more if we suspect what they say is true but don't want to face it ourselves. Most owners of older animals experience this response when someone who hasn't seen the dog for several years remarks, "Wow, Buddy's really showing his age, isn't he!" We feel the tiny hairs bristle at the back of our necks, and our stomachs tighten. "He is not!" we explode. Although such age-related comments are difficult to deal with, the feelings aroused when someone says something negative about our dog's behavior can be even more problematic. Consider these statements:

"I saw Morty gallivanting down by the highway yesterday. A couple of cars just barely missed hitting him."

"Are you sure Ginger's okay around kids? She looks a little skittish to me."

"You must have nerves of steel to put up with that barking all day."

Depending on the owner's interpretations of the dog's behavior, these may strike a raw nerve. Suddenly a previously ignored result of Morty's unsupervised romps rears its ugly head; he could get hurt or cause an accident on the busy road. We're forced to view Ginger's shyness, which always appealed to our protective instinct, as a source of possible danger to others. Enduring Freddy's incessant barking as a function of our patience and love for the dog is replaced by another's opinion that we're incredibly dense.

If others' interpretations either bother us or recur, we owe it to ourselves and our dogs to reevaluate our own orientation carefully. If we find ourselves becoming angry and defensive, chances are our own interpretations are the incorrect ones.

If we can't see anything wrong with our interpretation and don't feel that any change is necessary, then the best thing to do is to remove the dog from that environment. If we don't believe Rex is unruly, but the neighbors insist he is, keeping him on our own property and under our control whenever they're around resolves the conflict. If that's disagreeable, we can move, hope all our neighbors move and sell their homes to people who think Rex is as wonderful as we do, or accept living in a neighborhood plagued by dog-related strife for as long as Rex lives.

Thus we can see that accepting our own or someone else's interpretation of our dog's body language is a matter of choice. If Leigh is comfortable with her interpretation of Magoo's body language but can also accept her sister's evaluation of it, she can confidently embark on a program to reconcile her dog's behavior with both points of view. In Rollo's case, if the neighbor's interpretation of the meaning of the dog's extra pounds strikes Rollo's owner as more relevant than my medical evaluation, she'll also make any changes necessary to improve her relationship with her dog—and her neighbor.

In both of these cases the goal is a bonded relationship with the pet. But what happens when an owner's point of view is based on the furry-humanoid or wild-dog orientation? If owners view the dog's behavior anthropomorphically, their interpretation of the animal's emotions and body language will be inextricably intertwined with their own. For example, when Bit O'Honey cocks his head to one side just as Emily Sullivan prepares to leave the house, she relates that body language to her own thoughts at the moment. If she plans to be gone for a long time and feels guilty about not taking the Scottie, she may say, "You're right, Mommy *is* mean to go away and leave you!" and interpret the cocked head as an accusation. If she intends to take Bit O'Honey with her, she responds to that same body language with "But of course you can come with me. I wouldn't dream of leaving without you!" Instead of recognizing the body language and any precipitating emotion as a reflection of human and canine points of view, Emily interprets both interactions solely in terms of her own orientation. In contrast, Lou Rutherford views Merlin as a totally separate species whose body language reflects the differences between their two species. When the black Lab cocks his head, Lou almost invariably asks, "What do you hear, Boy?" When the black Lab cocks his head, he could be displaying attentiveness to someone or something within Lou's perceptual range—the sound of children playing outside, for example. Or Merlin might be signaling discomfort from a minor or not so minor ear infection. However, Lou almost always assumes that whatever his dog experiences is outside his comprehension and responds with the unanswerable question, "What do you hear, Boy?"

While neither orientation may be detrimental under normal circumstances, both can become quite perplexing when others' interpretations of our dogs' behaviors don't coincide with ours. Because Emily views Bit O'Honey as an extension of herself, to differ with her interpretation of the dog's body language is to differ with Emily herself. If Emily interprets certain body language of her dog to mean "I love you," because it seems similar to the behavior she uses to convey love to her husband, and her brother comments, "Bit's trying to tell you he needs to go out," when the dog displays that body language, Emily perceives the difference in interpretation as more than a difference in point of view. Because of her highly anthropomorphic relationship with Bit O'Honey, Emily interprets her brother's remarks to mean that *she* can't tell the difference between the body language expressing love and that indicating a need to urinate or defecate. Needless to say, when Emily becomes openly defensive and upset by his view, her brother is quite confused.

On the other hand, when someone offers a point of view that differs from Lou's, he sees this as increasing the distance between him and his dog even more. When Merlin howls and Lou informs his pals the dog is trumpeting a challenge to the mastiff next door, and one of his pals says, "I thought it was because of those sirens. They always make our dogs howl," Lou feels uneasy. Is his buddy right? Is his dog's hearing really sensitive to sirens?

In both cases the owners' orientation and resulting highly limited point of view makes it quite difficult for them to consider other views of their dog's body language objectively. Rather than taking these comments as input that might help them to get a different slant on problem behavior or expand their basic knowledge, they perceive any point of view that differs from their own as threatening to their relationship with their pets.

DECIDING TO SOLVE THE PROBLEM

Once we recognize a body-language or emotional problem, define it, and ascertain the point of view from which that definition springs, we're ready to consider possible solutions. To avoid being overwhelmed by needless details and complicated processes, it makes sense to deal with two fundamental issues first:

- Do I *want* to solve this problem?
- *Why* do I want to solve this problem?

Although these may sound like variations of the same tune, owners should answer both questions before undertaking any training program. For example, in the five years Leigh and Magoo have been together, Magoo has chewed old rags and dog toys when left alone. This never bothered Leigh until Magoo ruined a pair of her expensive shoes.

Problem definition: a dog who has exhibited chewing behavior in his owner's absence for almost five years. Because he's never damaged anything valuable, his owner has tolerated the body language: "I thought it was a release for Magoo when he was lonely, kind of like sucking his thumb," muses Leigh.

When Leigh asks herself whether or not she wants to alter Magoo's body language, she hesitates much longer than the toddler's parents whose beagle pup is urinating and defecating in the baby's room. If the chewing behavior in and of itself exasperated her, Leigh wouldn't have accepted it for almost five years. Because she's not sure whether she wants to solve the problem or not, Leigh turns to the second question. If she can't state any good reasons for altering Magoo's behavior, any training effort will probably fail. For example, if she had carelessly left the shoes near the dog's empty food dish, Leigh's incentive to begin a training program would be far less than if Magoo normally sleeps on her bed in her absence and had pulled the shoes from her doorless closet.

Suppose the shoes belonged to someone else? That could dramatically affect Leigh's decision about altering the behavior in one of two completely different ways. If she rarely entertains, she may decide simply to remember to warn visitors about Magoo's habit. If the incident created sufficient negative feelings—if the guest berated and embarrassed her terribly—Leigh might decide to go to any lengths to prevent a recurrence of the problem.

Just as all body language and emotion must serve some purpose in order to be sustained, so any alteration of those expressions must be backed by both logical and beneficial reasons. It won't do Leigh any good to initiate a training program to reverse a long-accepted behavior unless she believes that doing so will benefit her and her dog. Such belief depends, as it always must, on the situation and the individual traits of all parties involved.

CONSIDERING THE OPTIONS

Once we believe that some change must occur to enhance our relationship with our pet, we're ready to weigh possible solutions. Here again we want to strive for simplicity. Some owners define a problem, such as chewing, check several training books out of the library and then study all of the recommended ways to stop chewing behavior. Such a process may certainly produce results, but changing the behavior is only one of four potential solutions available, and we should consider all four. Should you:

- Accept the behavior and your feelings about it?
- Change your feelings about the behavior?
- Change your own or your dog's body language to align it with your feelings?
- Terminate the relationship?

Let's examine each option more closely. No matter what other people, experts or otherwise, may say, can you accept your dog's behavior, the emotions you believe it conveys, and your own emotional and body-language responses to it? If you can, why try to solve a problem that doesn't exist? We already noted that if Leigh didn't find the chewing incident particularly upsetting, she probably wouldn't succeed in stopping the behavior. If Leigh can honestly say, "Oh well, I seldom wore that pair of shoes," and forget it, she and Magoo don't really have a problem.

However, suppose the owner can accept the behavior but

believes the dog or other people can't? Suppose every time Magoo embarks on a chewing spree he winds up vomiting or going off feed. Or suppose the chewing drives Leigh's housemate crazy, even though the dog has never destroyed any of the latter's property. In those cases, even though Leigh can accept the behavior, she must decide whether she can accept the negative effects it has on her pet's health or her relationship with other people.

If we can't accept a form of body language, then someone or something must change. The question is, Who or what? If our dogs look healthy and behave in a way that bothers only us, we may find it simplest to change the way we *feel* about the behavior. We noted in chapter one how owners who anthropomorphize their pets often have difficulty dealing with such common intercanine behaviors as sniffing and territorial marking. If the dog doesn't zero in on your guest's armpits or crotch, and if the territorial marking occurs outdoors and away from prized plants, why not learn to accept this normal body language rather than attempt to change it? Can anything be more futile, unrewarding, and of questionable value than struggling to teach a dog not to sniff? Choosing to be irritated and repulsed by such normal behavior makes as much sense as choosing to be unhappy every time the wind blows.

Assuming we find the body language unacceptable and can't change our own emotional reactions to it, the third option involves changing the dog's and/or our own body language to align it with our beliefs. Notice how we include human changes and don't automatically assume that only the dog needs retraining. If necessary, humans can change, too; and a willingness to do so can speed the process of improving any relationship. Certainly if Leigh decides to spend the necessary time and money to hang a sturdy door on her closet, this change on her part will produce the desired results: Magoo can't chew shoes he can't reach. On the other hand, if Leigh wants to be able to trust Magoo not to chew regardless of the availability of objects, then she must consider changing her dog's behavior. In both cases, if she regards the underlying emotion leading to Magoo's behavior as negative—such as spite or meanness—she must also be willing to alter her erroneous view as part of the remedial process. Although we can accomplish body-language change without a corresponding change in emotion, doing so complicates the procedure.

When faced with a problem, most owners usually do decide to change their own or their dog's body language or behavior, not so much because doing so makes such good sense, but because doing so is probably the easiest route in the long run. While it may seem easier merely to accept the ruined shoes and forget the incident, most people find that a lot easier to say than do. Suddenly Leigh receives invitations to parties where those shoes would perfectly complement her outfit; and when she goes out, now she's plagued by vague fears about what Magoo may be getting into in her absence.

Furthermore, just changing our attitudes doesn't give us much to *do*. If our primary goal is not to feel what we were feeling before, we completely internalize the change and miss the support of rewarding interactions with our pets. Think of something your dog does that irritates you—barking at motorcycles, for example. Imagine yourself deciding not to let this bother you; it doesn't bother anyone else, and you have neither the time nor the desire to train him not to do it, nor can you afford to pay someone else to train him for you. How do you feel every time a motorcycle roars past your house, triggering a frenzied response? Does every cell in your body scream to respond negatively, both emotionally and physically, to the behavior as you have so many times in the past? Do you feel yourself cringing, gritting your teeth, frantically searching for something to distract you until the barking subsides?

Compare this to initiating a program to change your own or the dog's body language, a program that requires planning and constant vigilance. When the motorcycle roars by and you move to smack Lucifer, you force yourself to withdraw the rolled-up newspaper, put it down, and carefully remind yourself that the dog isn't acting out of spite or stupidity. You train yourself to be ready to tell Lucifer to sit *before* he starts responding to the motorcycle; you fix him with your gaze and hold his attention until the noise recedes. Then you lavishly praise his previously unrecognized brilliance. Grueling? Emotional? Perhaps; but when you let the newspaper drop and feel the anger dissipate, when you see Lucifer sit quietly, you have tangible proof that you and your pet are making progress.

TERMINATION: THE END OR A NEW BEGINNING?

Whenever I have posed this option, no matter how careful my words and regardless of the seriousness of the problem involved, I invariably shock people. However, because any good relationship deserves—no, *demands*—honesty above all, and because the idea of terminating a relationship occupies one end of the spectrum of choices owners confront, we must consider it. If we don't recognize this fact, we place no limits on problems; and problems with no limits can go on and on, causing unlimited physical and/or emotional pain for all involved.

Why should Leigh consider something so drastic as getting rid of Magoo—either giving him away or having him euthanized—because of one chewing incident? Remember, each expression of body language and emotion in an interaction between owner and dog is relative, and today's acceptable behavior may deteriorate into that which is intolerable tomorrow. Because dogs, people, and circumstances change, we owe it to ourselves and our pets to evaluate each problem not only in terms of present problems but in terms of our full range of available options for solving them. Failure to do so can allow seemingly insignificant infractions to pile up and destroy a relationship.

We already noted some circumstances that would lead Leigh to accept Magoo's behavior and others that would induce her to change her own or the dog's response. Now let's consider some circumstances where terminating the relationship might have merit. Obviously, if Leigh finds the chewing intolerable and can't or won't alter her opinion and/or Magoo's behavior, she can either perpetuate a negative relationship that can only get worse or terminate it. If her relationship with her housemate means more to her than anything in the world, she will also see getting rid of the dog as a viable option. Finally, if

Leigh's relationship with Magoo is already on shaky ground, if she's been looking for that last straw, one relatively minor chewing incident may provide reason enough to call it quits with Magoo.

Consideration of the fourth option can benefit dog owners who approach it. If we're afraid to consider termination, it will nonetheless rattle like a skeleton in the closet of our relationship, creating guilt, fear, and anger every time a fleeting hint of it crosses our minds. Compare the following ways Leigh might react to this option:

"It's barbaric; I can't even think about such a horrible thing!"

"I've given it careful thought and decided I will never get rid of Magoo."

Which orientation better prepares Leigh either to accept or to alter her relationship with her dog? If she squarely confronts the notion of termination, considering all the pros and cons, and decides against it, she's free to channel all her energy toward accepting or altering the behavior. By refusing to consider termination, part of her energy must invariably stray in that direction, exerting negative and counterproductive effects on any other course she may choose. Not considering termination is like having a tiny hole in the bottom of a bucket and wondering why the bucket's never full. Only by acknowledging the existence of the option of termination and dealing with it honestly can we free ourselves to experience our pets to their fullest.

What if our consideration leads *toward* termination? Can the loss of a healthy pet ever be considered beneficial? In order to answer this question it is often helpful to ask, What is the worst thing that could possibly happen? Leigh could answer, "I get rid of Magoo to please my roommate and then discover something else is bothering her." Or she might say, "I get rid of Magoo and discover how much I really need him."

In the first case, Leigh realizes that the actual problem may not be her relationship with Magoo at all, but rather that with her housemate. If so, getting rid of the dog serves no purpose because the dog isn't the problem. In the second situation, Leigh opens herself to the possibility that good qualities about her relationship with Magoo may have been masked by her frustration and anger regarding his misbehavior. Either way, her honest confrontation of the worst possible results of termination give her new insight that may actually help her preserve their relationship.

If attempting to fool ourselves about the possibility of termination as a viable alternative causes problems, withholding our views of this option from others intimately associated with the dog can cause even bigger problems. I learned this lesson from ten-year-old Tim the day his beloved Dalmatian was struck by a car and severely injured. Tim's parents told him to stay in the waiting room while I examined the dog; as soon as he departed, they begged me to do everything in my power to keep the dog alive for twenty-four hours. Tim was scheduled for major surgery the next day, and they were terrified of upsetting him.

"Lie, if you have to, but tell him Smokey'll be alright," his sobbing mother implored.

I was torn between my revulsion at lying to a child and my compassion for the parents' anguish. As I walked toward Tim, I prayed for inspiration. Tim stopped playing with his ball, looked me in the eye and said, "Smokey's going to die, I know she is. And Mom and Dad don't want me to know. It's alright. If she doesn't suffer, I'll be okay too. I want you to put her to sleep." He struggled to control his emotions as only a ten-year-old boy who's just made a very grown-up choice can. I hugged him and turned away quickly so that I wouldn't embarrass him by seeing his tears.

His parents rushed up to me as I reentered the examining room. "Did he believe you?" They anxiously searched my face for the answer.

"Tim knows Smokey's going to die, and he wants me to make sure she doesn't suffer. If you agree, I suggest we put the dog down immediately."

Tim's surgery was uneventful, his recovery rapid and complete. Maybe that would have happened anyway, but the rapport this child had with Smokey and the setter he got several months later convinced me that withholding any information about his dog, regardless how bad, would be the worst thing any truly caring person could possibly do.

When parents don't raise the termination option and discuss it openly, their children often think that their parents are avoiding talking about it. When Norton repeatedly messes in the house and the elder Petersens scream and holler and smack the dog, why wouldn't the children believe that their parents hate the pup and might want to get rid of it? Yet time after time parents are stunned to discover their children ever harbored such fears.

Sons and daughters whose elderly parents' pets develop medical or behavior problems also do their elders no favor by evading this option. Oftentimes offspring feel that an open discussion of canine euthanasia or giving the dog away gives tacit approval for doing the same with humans. Maybe some individuals do anthropomorphize their pets and imbue them so thoroughly with their own characteristics and fears that such a connection would be made; however, in my experience this occurs rarely. More often than not, elderly owners want what all pet owners want: a quality life for their pets and a quality relationship with them. If that quality isn't available, then termination becomes a compassionate alternative.

When octogenarian John Finkelstein's fifteen-year-old pug lay critically ill in the veterinary hospital, John was beside himself because the reports his daughter relayed were so vague. Every time he asked, "Is Eloise going to be alright?" or "Is Eloise going to die?" his daughter or son-in-law patted his arm reassuringly and told him not to worry. Finally John became so frustrated that he waited until his family left the house and called the veterinarian himself. He learned that Eloise lay in a coma, totally dependent on external support systems and with little chance of recovery. Although John felt heartbroken, he wasn't surprised because he knew his dog so well that he had sensed the end coming weeks ago. He confronted his daughter and son-in-law with the information, telling them he found such extraordinary treatment unacceptable for himself and certainly for his dog. He asked that they respect his decision to have her euthanized.

The option to get rid of the dog isn't a dollars-and-cents decision. We're not talking about money, time, or effort; we're talking about the quality of a relationship between two living beings. The fourth option is the acid test, the Rubicon that will either give us the necessary energy to implement changes or force us to face the reality of a less-than-perfect relationship if such changes aren't possible. Don't try to sneak past this option with the false hope that it doesn't apply to Mugsy's blanket sucking; if the sucking bothers you or creates problems, you must consider this option along with all the others.

Dog owners don't encounter problems from making wrong decisions nearly as often as they do from making no decisions at all. Before initiating any training program, carefully evaluate all four options and consciously choose the best for you and your dog.

Did anything you read in this chapter make you feel guilty? In the next chapter we're going to take a look at guilt and see how we can come to grips with this nightmare emotion in our relationships with our pets.

4

GUILT: THE NIGHTMARE EMOTION

LOU RUTHERFORD PRIZES only one possession over his black Lab, Merlin: an outfielder's glove autographed by the great Boston Red Sox slugger Ted Williams. He usually keeps the mitt in a display case with his high school football helmet, his bowling and softball trophies, and a collection of antique hand guns. However, one evening after his team won the All-City Softball Championship, Lou was in such a hurry to shower and get to the celebration at the Left-Field Tavern that he left the glove on the coffee table near Merlin's favorite napping spot. When Lou returned home after more than his usual number of boilermakers, he found the shredded glove in the middle of the floor. Enraged, he bellowed and lunged at the dog. Merlin froze, ears back, tail tucked tightly against his abdomen. Seeing his dog so obviously aware of his misbehavior, something snapped in Lou. Grabbing a hockey stick, he struck the dog again and again. Never had he beaten Merlin so badly.

That night Lou tossed in his sleep, plagued by a vivid nightmare in which a massive wolf chased him through a maze of brambles and briars before finally pinning him down. Waking in a cold sweat, images of Merlin's guilty expression and the shredded glove melted into those of himself beating his pet and of the terrorized look in the dog's eyes.

Have you ever awakened from a nightmare in which you were relentlessly chased and threatened, in which you stumbled and fell, and couldn't catch your breath or see where you were going? When you awoke did your heart pound, and did it seem to take ages before you realized you were alright, that it was just a bad dream? Then, despite the reassurance of your familiar surroundings, did you still feel a lingering fear in your heart?

Many times guilt works the same way. It sneaks in and violates a beautiful relationship, much as the odor of skunk can mar the most breathtaking scene, leaving us with a sense of uneasiness and irritability long after the actual details of the event have disappeared from memory. When a sobor Lou Rutherford awakens in the morning and remembers beating Merlin, he senses that his relationship with his dog has changed in a way that may permanently damage what had been until then a source of great pleasure for them both.

SINS OF OMISSION AND COMMISSION

We can invariably break down what we consider wrong behavior in ourselves and our dogs into two basic categories:

- Sins of omission—those acts we or our dogs should perform but don't.
- Sins of commission—those acts we or our dogs do commit but shouldn't.

For example, if Merlin should come when his owner calls him but doesn't, or if Lou should have Merlin vaccinated every year but forgets, both have committed sins of omission. Lou evaluates both behaviors as wrong because he and Merlin have failed to exhibit preferred behavior in particular situations. If Merlin chases a car or Lou hollers at his dog simply because he's in a bad mood, Lou can classify these behaviors as sins of commission.

Anytime anyone or anything commits a sin of either type, both the perceiver and the performer of the act can succumb to feelings of guilt. When Lou sees the shredded glove, he defines this as a major wrongdoing for which he expects Merlin to feel guilty. Whether Merlin exhibits true signs of guilt or not, chances are Lou expects him to do so; nor is Lou surprised that he feels guilt when he recalls the subsequent beating. Regardless of whether dogs experience what humans call guilt, most owners can cite numerous instances they believe demonstrate their dogs' awareness of some transgression. And almost every owner can confess to at least one case of losing patience and doing something to the dog for which he or she still feels a pang of guilt.

Although owners may relate guilt in themselves and their dogs to specific *acts*, guilt really reflects more on the *relationship* between human and animals. To discover more about how guilt works, let's take a closer look at this emotion and its related human and canine body language.

GUILTY CANINE SIGNALS

Consider this list of some of the most common body-language signals people associate with guilt in their dogs:

- Averted gaze.
- Trembling.
- Refusal to move.
- Hiding.
- Refusal to come when called.
- Refusal to eat.
- Rolling over and exposing the abdomen.
- Flattening the body against the floor.

As we examine this body language, we quickly see that much of it expresses submission. For example, dogs who won't move, come when called, flatten themselves against the floor, or hide under the bed all exhibit the freeze response of a submissive animal in the presence of a perceived dominant individual or force. But do these signals also express guilt? Whenever Emily Sullivan comes home, Bit O'Honey normally jumps up to be petted. After dinner, the Scottie joins her in the living room and rolls

over to have his belly scratched. But why does the dog sometimes roll over the instant Emily comes into the house? "That's
simple," says Emily. "He only does that if he's gotten into the
trash. He *knows* when he's been a bad dog." Like beauty, guilt
often depends on the eye of the beholder.

I remember seeing an obviously terrified dog whose owner
finally brought him to the clinic after a thorough investigation
of her household failed to produce any destruction to account
for the dog's guilty expression. In this case, a complete examina-
tion failed to reveal any physical abnormalities. However, as the
owner detailed changes in the dog's environment during the
preceding eight hours, it seemed likely that the mild Bedlington
terrier had fallen victim to that peculiarly northern disease,
Snow Slide. Imagine yourself a shy, twenty-three-pound dog
left in charge of a large house. Three feet of wet snow has de-
scended during the night, but today the sun has returned to
warm the world. Just as you're dozing off in your favorite patch
of sunlight in front of the living-room window, your hear a faint
creak. You perk up your ears. *Crack!* As you rush frantically
from window to window, the sound seems to come from every
direction at once. You can't escape it, nor can you determine its
cause. To your sensitive ears it sounds like thundering doom.

Most New Englanders rejoice when Mother Nature frees
them from the labor of shoveling heavy snow from the roof; on
the other hand, more than one dog and more than a few humans
have been scared out of their wits by this phenomenon. I've also
traced the causes of other so-called guilty expressions in dogs to
slipped discs, arthritis, and loss of vision or hearing. In these
cases, what the owner perceived as guilt the veterinarian interpreted as simple anxiety and fear.

A MATTER OF DEFINITION

From such examples we can see that circumstances, relationships, and individual idiosyncrasies play a major role in the
interpretation of submissive body language as an expression of
guilt. If Merlin has exhibited destructive behavior over the
years, Lou will automatically look for evidence supporting this

belief whenever Merlin exhibits submissive behavior "for no good reason." Because Bit O'Honey normally behaves so well, when Emily Sullivan detects submissive body language in her dog, she's more likely to assume *she's* done something wrong: "Mommy buy Bitty the wrong food?" "Mommy gone too long?" If the owner feels confident about his or her relationship with the dog, this same body language could simply indicate that something is bothering the dog, but not necessarily a wrong "something." "Hey Magoo, not feeling up to snuff today?" Finally, when veterinarians, trainers, or other people who don't know the dog intimately view such body language, they may interpret it quite differently or attach no significant meaning to it at all.

Once we define "that look" as meaningful body language, we must determine exactly what it means to us and decide whether we can accept it or not. As we noted before, whether dogs actually feel guilt or not matters less than the fact that the overwhelming majority of owners believe they do. Although behaviorists could argue that guilt produces no benefits for wild dogs beyond those obtained through submission and that therefore any guilt is a purely human fabrication, the fact remains that guilt really does affect most human/canine relationships.

Obviously if guilt exists in a relationship, it must serve some purpose: Why would a dog feel and act guilty? "That's simple," claims Emily. "The dog loves its owners so much it can't bear the thought that it's done something to make them unhappy. That's why they get that guilty look or go belly-up."

"Horse baloney!" snorts Lou. "A well-trained dog knows damn well when it's done something wrong, and it knows it's going to be punished. *That's* what makes it look guilty—fear of well-deserved discipline."

These two statements represent two of the most common ways owners interpret the submissive body language their dogs direct toward them. In both cases the dog's submissive response to the owner's anger or other dominant cues becomes erroneously linked to whatever sin the owner perceives the dog to have committed. When Lou hollered at eight-week-old Merlin for chewing an old sneaker, what really incensed Lou was that the pup committed this transgression without guilt, or rather without what Lou considered appropriately guilty body language. "You should have seen him wagging his tail and, I swear, even grinning. I knew right then and there I had to show him who was boss or I'd never be able to control him!"

Although Merlin chewed the next object with equal relish, he quickly evaluated Lou's dominant body-language response to the act (brandishing a whistling willow switch) and assumed the corresponding submissive stance—whereupon Lou punished him anyway, teaching Merlin that even more submission was required to atone for the sin. If Merlin hadn't displayed submissive signs, Lou would undoubtedly have seen defiance in the dog and erupted with louder shouting and a more severe smacking. Either way the owner has created and perpetuated the "guilty" behavior.

When we define canine submissive displays as signs of guilt, we set an intricate and diabolical trap for ourselves. If we're convinced these expressions mean the dog has done something wrong, then in order to preserve our definition, we must find something wrong anytime the dog displays that body language. For example, when Lou inadvertently startles a distracted Merlin and the Lab responds submissively, by his own definition Lou must find something his dog has done wrong. "Merlin, what were you up to?" he demands sternly. The dog cowers in response to Lou's voice. Now Lou has created a no-win situation for his relationship with Merlin. If he can't discover any canine sins, that means his definition of Merlin's expression is incorrect; and, like most humans, Lou doesn't like to be wrong. On the other hand, if Merlin actually did do something wrong but Lou can't find it, that means Lou's been outsmarted by his own dog, an equally unacceptable situation. Lou *must* find something wrong, and as all dog owners know, this is usually easy. There's always a stain, a ding, or a snag we can blame on the dog. So what if Lou winds up disciplining Merlin for a real or imagined sin that occurred weeks ago? Lou may be able to maintain his belief that "that look" always signifies guilt, but the dog cannot possibly associate today's punishment with last month's transgression.

In such ways we literally train our dogs to respond submissively to what we consider negative behavior, oblivious to the fact that we're training them to respond submissively to us and not guiltily to the behavior. Once we achieve a consistent submissive response, we define it as guilt, then respond in whatever ways we deem necessary to perpetuate our definition. So even though we may believe that we are linking up our dogs' guilty submissive body language with some specific negative act, in truth we are producing more of a problem with our relationship to our pets than with their behavior.

I'M GUILTY, WE'RE GUILTY, YOU'RE GUILTY

Once we recognize guilty body language in our dogs, the ball's back on the human side of the court. How do owners respond to submissive signals they perceive as an expression of guilt? Depending on the person's relationship with the pet and the circumstances, these signals may send several different messages: "I'm guilty." "We're guilty." "You're guilty."

If an owner views a canine sin and a subsequent submissive look as a dog error, the owner will probably see submission as a sign that "I'm guilty." For example, if Emily is busily cleaning house and catches Bit O'Honey lifting his leg against the laundry basket, yells at him, and sees him cower, she can comfortably interpret his submission as his way of saying, "I'm guilty. I know I shouldn't mess in the house and make more work for you." On the other hand, if one of Emily's friends brings a large aggressive dog to visit and the guest frightens the Scottie so much that after its departure Bit O'Honey manifests the same urinating display, Emily might see the behavior in a different light. If she feels responsible for allowing the strange dog in the house, but nonetheless finds her dog's marking repulsive, she sees her own guilt as well as his reflected in her dog's body language: "Mommy's sorry for letting that mean old dog in the house, but that's no excuse for your making a mess." In a third scenario, Bit O'Honey lifts his leg against the laundry basket, only to have Emily discover it when she returns from a long day of fun at the beach. Now when Bit O'Honey cowers in response to the tone of Emily's voice when she discovers the mess, Emily's own guilt at abandoning her beloved pet while she had such a grand time colors her interpretations of both Bit O'Honey's transgression and the meaning of his body language. "Poor baby missed Mommy and had to go so bad. I'm sorry, Bitty. It's Mommy's fault."

Let's examine another clue in the case of the great guilt trip. We begin with the dog's submissive display, which the owner perceives as a guilty response to something the dog did or didn't do. If we perceive the dog's guilty expression and what we associate it with as legitimately connected and respond in what we consider an appropriate fashion, the interaction ends there, and we experience no guilt ourselves. If Lou catches Merlin in the act of chewing his mitt, he reads the dog's submission as "Oh, oh, you caught me doing a bad thing; I deserve to be punished." When Merlin displays further submission in response to Lou's discipline, Lou interprets this as his dog's way of saying, "I'm sorry. I learned my lesson." Lou accepts these interpretations, forgives Merlin, and forgets the incident.

However, what happens if we feel that either our interpretations of canine behavior or our own responses to it are inappropriate? When that's the case, owners buy a ticket for the guilt trip too. If we feel guilty, then something about the interaction strikes us as incongruent:

- The dog's guilty expression wasn't related to the transgression.
- Our punishment was inappropriate or excessive.
- Our response to the guilty look and the act sprang from some other, perhaps evil, motive.
- We feel responsible for the negative behavior (that is, we punished the dog for something we could have prevented).

The first guilt-producing situation frequently occurs when owners discover their pets' guilty expression results from a medical rather than a behavioral problem. One woman client associated submissive body language in her dachshund with the animal's failure to eat its food, so she spent several days forcing her pet to eat very rich table scraps. You can imagine how badly she felt when I diagnosed a severe digestive disorder. The dachshund wasn't displaying guilt, it was displaying pain and anxiety; every time it was forced to digest more food, the pain and anxiety increased. Suddenly a look the owner previously took to mean, "I'm guilty for not eating my wonderful food and for making you unhappy," demanded a radically different interpretation. I had to get her to accept my interpretation of the look—"Something's bothering me"—rather than her own guilt-laden, self-punishing one: "Why do you keep hurting me? Don't you love me?"

If we view our reaction to the dog as excessive, we may also see our own guilt in its submission. When Lou loses his temper and beats Merlin, he becomes haunted by the memory of the dog's rigid, trembling body and terrified eyes. In this case, Lou interprets the initial submission as "I'm a very bad dog. I chewed your glove and know you will punish me." If Lou believes he disciplined his dog too harshly, he may view any submissive display during or following the beating as "Please stop. You're hurting me!"

If we lash out at our dogs because we're in a bad mood, their submissive response almost invariably makes us feel guilty. Have you ever been strolling along absentmindedly only to stumble over your dog sprawled on the floor? Did the shock or your subsequent fall cause you to shout, "Dang it, do you have to take up the whole room!" In such situations our own surprise and irritation can cause us to yell at our dogs and even discipline them. However, the instant the dog assumes that familiar submissive posture, we feel terrible. After all, Samantha was sleeping beside the green chair where she always sleeps; it isn't her fault you were daydreaming about the birthday party you're throwing for your daughter tomorrow.

The final form of owner guilt arises when the dog does something we foresaw but made no attempt to prevent. Lou knew Merlin was a chewer, yet he left his prized glove lying within the dog's reach. Lou meant to cart off those old boards and barbed wire, but didn't; true, Merlin wasn't supposed to play in that corner of the yard, but Lou naturally read accusation rather than guilt in the dog's expression when the Lab held up his bloody paw.

GUILTY OWNER RESPONSES: ATTENTION AND ALIENATION

Owners who believe they've misinterpreted their dog's body language and/or responded inappropriately manifest their guilt in one of two ways. Either they seek to decrease what they view as a gap between them and their pets by paying more attention to them, or they separate themselves from the animals in order to free themselves from a constant reminder of their inadequacies.

For example, when Emily punishes Bit O'Honey for urinating in the house after the strange dog visited them, she experiences postdiscipline guilt because she believes her dog would not have misbehaved if she hadn't invited the threatening creature. Therefore, she attempts to atone for her insensitivity by becoming intensely sensitive and solicitous of Bit O'Honey's needs. She may give him an extra helping of ice cream or three cookies instead of his usual two that evening. Other owners may suddenly decide they must rush to the store—"Peaches want to come for a ride with Daddy?" Still others may feel an inexplicable urge to make pizza—the dog's favorite treat—for dinner that evening. Even usually sedentary owners may, following a negative interaction with their pets, take the dog on an especially long walk through its favorite haunts: Maybe if they and Treavor spent more time together the negative behavior wouldn't happen.

If owners feel responsible for the dog's behavior or any excessive or erroneous reaction to it, they often believe that paying more attention to the dog or otherwise growing closer to it will prevent the negative interaction from recurring. Such owners gear their body language toward decreasing the perceived gap between them and their pets by becoming physically closer to them via treats, long walks and rides, or unusually lavish displays of affection.

Decreasing the distance usually appeals more to dependent owners who see their pets as extensions of themselves or in situations where the owner strongly identifies with the dog's experience. If the interaction makes Emily unhappy, she's sure Bit O'Honey feels unhappy too. When Merlin holds up his wire-enmeshed bloody paw, Lou remembers how angry he became when his wife left a rake lying in the yard and he stepped on it. Because Lou believes such negative feelings are justifiable but he doesn't want Merlin to be angry, he rushes to attend to and console his dog. In both cases the owners don't like feeling alienated from their pets and will go great lengths to reestablish a stable relationship.

When Lou views Merlin as a separate and sometimes even antagonistic species, accepting responsibility for misinterpreting his dog's body language or his inappropriate reaction to it creates problems. Does the fact that the interaction occurred prove that Lou lacks knowledge of his own pet or, worse, that he's the kind of guy who takes out his frustrations on a dumb animal? Either way Lou winds up seeing himself as an inferior human being. To remain in the company of his dog simply reinforces these negative beliefs, so he flees.

Impulsive owners like Lou may storm out of the house and stay away all day; more passive ones may stay home but attempt to ignore the dog. However, most of us find it harder to maintain isolation than increased attention because we're accustomed to responding to our pets and want to reestablish a stable relationship. No matter how hard we try to ignore Mopsy after she misbehaves, we soon find ourselves reaching for her favorite ball when she links up with that forlorn look on her face. Furthermore, it's impossible to alienate our dogs without alienating ourselves. Unless we can fully distract ourselves with other activities, we either abandon this approach or become frustrated and angry with our inability to maintain it. If we can abandon it without feeling we're giving in to the dog, the episode terminates without permanently damaging the relationship. However, if the separation or its termination leads to anger, the relationship undoubtedly suffers. For example, imagine that after Merlin destroys the glove, Lou smacks the dog repeatedly, then storms out of the house to "cool off." As he drives aimlessly around, a number of thoughts race through his mind:

"I had to get out of there, or I would have killed him."
"He'd better not touch anything while I'm gone, or he's really going to get it!"
"Why the hell am I driving around when I've got so much work to do at home?"
"Why did I ever get that stupid dog in the first place?"

If Lou returns home at this point, chances are he'll blame Merlin for "driving me out of my own home" in addition to chewing the mitt. On the other hand, if Lou buys a new baseball glove and drives over to his best friend's house for an afternoon of batting practice and pleasant companionship, his view of Merlin's behavior will most likely improve by the time he returns home. Having dissipated his own negative feelings through positive activities, Lou's physical tiredness and good feelings should enable him to relate neutrally, if not positively, to his dog when they meet again. However, this can occur only if Lou separates himself from Merlin with the idea that he wants to dissipate his anger and guilt and forget about the destroyed mitt. If he goes off and broods about Merlin's sins, the separation can actually intensify his negative feelings.

Whether guilt-ridden owners use attention or alienation to dissipate their guilt, they share the same goal: reestablishing a stable, guilt-free relationship with their pets. Sometimes it works, sometimes it doesn't. We saw how a separation characterized by brooding undermines a shaky relationship even more. Similarly, if Emily fawns over Bit O'Honey to assuage her own guilt when, in fact, the dog has done something that deserves discipline, she nullifies any positive effects of the interaction. Nothing frustrates a veterinarian, trainer, or groomer more than to watch an owner whose dog has just attempted to remove one's ear, smack the dog and then apologize to it. Regardless of how guilty owners may feel about their dogs' behavior, their culpability, or their responses, the fact remains that the animal has performed a truly antisocial and negative act. A conscientious and loving owner should deal with such negative behavior when it occurs; to discipline the animal and then immediately try to soothe it merely makes a bad situation worse.

Once again we see how guilt and its impact on a relationship between owner and dog can cause more problems than the misbehavior itself. Therefore, before we can solve any problem, we must learn to understand the nature of guilt.

GOOD GUILT, BAD GUILT

If guilt does nothing to solve the initial problem, why don't we all simply eliminate guilt from our relationships with our dogs? As we noted, despite its nightmarish qualities, guilt can appear to offer redeeming qualities under certain circumstances:

"If I didn't feel so guilty about Tippy's weight, I'd never withhold extra treats."

"If I didn't feel so bad about Skeeter getting hit by that car, I'd let her run loose."

"I felt like such an irresponsible jerk after Pookie's surgery, I'll feed him that special food for the rest of his life."

"It was my fault Ernie got Parvo because I didn't get him vaccinated. Now he's going to get the best possible veterinary care."

Although we might accept such guilt as positive from a medical point of view because it leads to healthy changes for the dog, what ultimate effect does it have on the relationship? How long can we sustain behavior, regardless how beneficial its short-range effect, if we sustain it only to atone for our sins? If Skeeter recovers totally from her injuries and barks and chews frantically when confined, her guilt-ridden and now exasperated owners must ask themselves just how many weeks of canine confinement it takes to assuage their guilt. Some owners may think three weeks sufficient; others may feel a need for self-punishment lasting the dog's entire lifetime. In either case, the relationship between owner and dog can never be ideal. Not only haven't the owners solved the initial car-chasing problem that led to the accident, they now own a dog that also barks and chews. If they turn her loose, they'll feel doubly guilty if anything happens to her; if they keep her confined, her negative behavior will eventually undermine what was once a good relationship.

While it may be effective to use guilt to energize short-term changes, such as ensuring that Skeeter gets her medication every day during her two-week recovery, relying on guilt rarely accomplishes lasting change. As the examples indicate, it's not uncommon to use guilt to achieve desired results in medicine. If owners of sick dogs don't employ it themselves, their veterinarians often will, adding small doses of guilt to their prescriptions:

> "You wouldn't want Snickers's hips to give out because of all that weight, would you?"
> "Be sure Hermie gets all his puppy vaccinations or he could get very sick and die from something you could have prevented."
> "This medication will do the trick—as long as Topper gets it every eight hours."

Groomers and trainers may also use guilt to get owners to do things they consider beneficial to the dog's well-being.

If others attempt to use guilt to get owners to do something perceived as necessary to precipitate changes in behaviors that threaten canine life, human safety, or the dog's well-being, is such guilt "good guilt"? If owners consistently ignore life-threatening or antisocial behaviors in their pets, is it the duty and responsibility of others to create sufficient owner guilt to rectify the problem? Although this widely accepted view is often touted as a moral obligation, I personally can't accept it: It's the moral obligation of those with knowledge to teach, not to evoke guilt.

To be sure, I've seen owners awash in guilt over avoidable canine behavior; I've spent hours mulling over disastrous canine and owner behavior I foresaw but was unable to prevent because I couldn't convince the owner to make a necessary change; and I've lost sleep because I was incapable of creating and reinforcing change in an animal I only saw a few times a year. That's the crucial point: No one but those committed to the relationship can really initiate and reinforce lasting change. Trying to convince others to change their own or their dog's behavior based on guilt is like building a house on quicksand.

I remember one case in which the owner of a gorgeous pointer made subtle references to her dog's aggressive behavior at home. Because the dog acted so beautifully in the animal hospital under less than optimal circumstances, I initially chalked up the comments to owner insecurity and recommended she spend more time working with the dog, teaching him simple commands. However, the implications kept creeping into her conversations: "You know, Hondo is a great dog as long as we don't bother him while he's eating." "Hondo's impeccably well mannered, but don't mess with him while he's napping on the couch." I not only suggested consistent daily training to establish a more equitable relationship between human and canine, I proposed simple ways to alter the specific behavioral problems. However, the owner ignored my advice and refused even to consider hiring a professional trainer until her four-year-old son underwent six hours of facial surgery to repair the damage done by the dog. However, by then all the family members were convinced Hondo was "too far gone." When I euthanized the dog several days after the incident, I shared the owners' sadness and hoped that all had learned a valuable lesson that would serve them well if and when they got another dog. "Oh no," lamented the injured boy's mother, "We feel so guilty about what happened, we'll never get another dog!" Although it bothered me that her children would never experience the benefits of a sound relationship with a pet, how could I recommend getting another dog after all the family had suffered?

Four months after I euthanized Hondo the family surprised me by getting a setter pup. "Tommy wanted another dog and after what happened, we can't deny him anything," confided Tommy's still-guilt-ridden mother. At six months of age the dog began displaying all the negative characteristics of the pointer, and the owner insisted I put it down immediately. Everyone felt guilty, but no one had been willing to change.

The more serious the problem and its ramifications, the more owners, veterinarians, trainers, and handlers owe it to themselves and others to divorce guilt from any resolution. Because our perception of guilt in the dog arises from our relationship with the animal rather than the animal's misbehavior itself, concentrating on or enhancing the guilt serves no purpose beyond undermining the relationship further. Although we can benefit by becoming aware of the role guilt plays in how our dogs react to us and we to them, using guilt as a reason to initiate long-term training to stop biting, car chasing, or chewing gives us nothing positive to work with. When our guilt regarding our dog's or our own misbehavior provides our incentive, we begin with the basic premise that we and/or our dogs are wrong, outlaws, and misfits. This works if we really do believe that we and our dogs are antisocial and evil. But more often than not, we don't really accept that; we see ourselves as misguided and perhaps naïve, and our dogs as untrained or inexperienced. Viewing ourselves and our dogs as wrong clashes with this latter evaluation. Invariably we resist the recommended changes, not because we disagree with them, not because we *want* a dog that bites or chases cars, but because guilt has immobilized us.

Whenever owners, veterinarians, trainers, groomers, or anyone else evokes guilt as a reason to change human or canine behavior, they're reflecting their belief that the owners don't want to do what's best for their dogs. If we need either our own guilt or that contributed by another to make us do something for our dogs, we and our dogs are suffering something far more serious than a medical, behavioral, or grooming problem. We're trapped with our dogs in a fundamentally unsound relationship, which may in fact be creating all those medical, behavioral, and grooming problems.

DEALING WITH GUILT

Embroiled in these complex exchanges of body language and emotions, should we be dismayed that they contribute absolutely nothing to either the definition or the solution of the perceived initiating problem? Having lavished cookies and ice cream on Bit O'Honey to eliminate her guilt, Emily has achieved nothing but a fat pet. Having driven around for hours or having spent the day at his friend's, Lou has either burned up half a tank of gas or improved his pitching arm. Neither owner has done anything to accept or change the misbehavior they associate with their dog's and their own guilty feelings and body-language expressions.

Whenever guilt infects a relationship, the only cure comes from confronting it. Whether we think the dog's guilt stems from its natural behavior or from its anticipation of punishment, whether we think our own guilt results from our response to the dog's behavior or our own culpability, we can't enhance our relationships with our pets unless we recognize this emotion for what it is and deal with it honestly.

Guilt is the nightmare emotion because it entangles us and our dogs in a bitter chain reaction of body-language and emotional responses that ignore the underlying problems. In fact, guilt almost always distances us further and further from the heart of the matter. Therefore, the only way we can free ourselves to resolve the "real" problems—the chewing, biting, car chasing—is to deal with the guilt *first*.

How do we start? We can reduce potential solutions to four basic possibilities. First, we can accept the guilt and how it makes us feel about our pets and ourselves. Emily can recognize her own and Bit O'Honey's guilt-induced cookie binges as a normal part of their relationship. Of course, doing so implies that she accepts any additional sorrow and guilt that may arise as her dog's slender silhouette enlarges. Lou can accept his confrontations with Merlin and their effects as normal and learn to endure them, although again he may have to live with the fact that his violent approach upsets him and may someday badly injure his dog.

Acceptance is the most common approach chosen by owners who blame themselves for the initiating behavior and therefore want to punish themselves as well as the dog, replacing physical self-flogging with mental and emotional self-flagellation. Although when it is worded this way, we might wonder what kind of demented person would choose such an approach, many loving owners do opt for this route. "I don't like to see Bit O'Honey look like that," admits Emily. "But it does let me know he's done something wrong or I've upset him. I don't like the way I feel after I've punished him either, but the memory of that look makes me try to be a better owner." Lou Rutherford states different reasons for accepting his dog's submissive body language as guilt-based: "When I see Merlin look like that, I know he *knows* he's done wrong. I can deal with a dog that makes occasional mistakes, but I can't tolerate one that's so stupid it can't tell the difference between right and wrong, or so defiant it doesn't care." Relative to his own guilt feelings regarding his relationship with his dog and how he expresses them, Lou notes, "If I didn't have the memory of the terror in Merlin's eyes or that sick feeling I get when I pound him, I'm sure there are times I'd kill him. My guilt keeps me from getting so angry I hurt him. He's my pal."

The second choice we have is to accept guilt as a normal part of our relationships with our pets, but change any negative feelings we have about it. For example, suppose Emily and Lou decide their own guilt serves no useful purpose. Rather than serving as a canine indicator and a beneficial self-reminder, suppose their own guilty feelings make them feel that their dog's submissive behavior is somehow *forcing* them to feel or do something they wouldn't otherwise feel or do. Now, in addition to feeling guilty, they also feel manipulated, which, in turn, leads to resentment. If we can remove these latter feelings rather than simply endure them, the guilt becomes more manageable. For example, if Emily can view Bit O'Honey's submissive displays and her own postdiscipline cookie indulgences as normal expressions of guilt and let it go at that, she can avoid any subsequent resentment.

However, as we noted in chapter three, just changing how we feel about something doesn't give us a whole lot to do; unless we find some way of distracting ourselves from the negative feelings that occur when we interpret our dog's body language as guilt-based or accusatory, this option rarely offers a permanent solution. In essence, we simply learn to accept the previously unacceptable gracefully; if we feel this has been forced on us or occurs because "we have no other choice," we'll simply add more negative emotions to the relationship.

The third option involves altering the behavior we think relates to the guilt. Bear in mind that the body language directly related to the dog's guilt isn't that which resulted in the barking or chewing; it's the subsequent displays most behaviorists associate with submission and fear. Therefore, in order to implement this solution, we must alter one of our pet's most basic and intuitive patterns of behavior—not unlike teaching a right-handed adult to write with his or her left hand.

On the other hand, this approach can offer some benefits. Although the submissive display is certainly innate in the dog, dogs wouldn't exhibit that behavior if others didn't reinforce it. We noted how Lou initially disciplined Merlin following the first chewing incident because the Lab didn't act guilty; that is, he didn't exhibit submission. From that day on, Lou's anger and related smacks and shouts became linked with Merlin's submissive body language; the dog was responding submissively to his owner, not guiltily to his own misbehavior. That being the case, if we can alter the relationship between owner and dog such that these submissive displays aren't necessary, we can free the owner from associated guilt.

The major drawback with this approach is that changing our relationships with our pets almost invariably means we must be willing to change ourselves first. Anyone who's sat for hours trying to coax a terrified dog out from under a bed knows the difficulty of convincing a dog to treat a previously threatening individual as its best friend. However, it can be done. Many times owners of shy young canines are horrified to discover their pets urinate submissively at the slightest provocation. Because they don't realize that the animal simply wishes to show respect or fear, they discipline it; but doing so merely increases the dog's submissive behavior. Then, if owners associate the submissive signals with the sin of urination, they attribute these signs to their animal's guilt. When these owners discover the true meaning of their pet's behavior, most become willing to alter their own. Instead of yelling at the dog or otherwise inducing the submissive response via their own deliberate or inadvertent dominant behavior, they learn to ignore the dog's submission and eventually see the response extinguish itself.

Of course, if we feel that altering our own behavior to eliminate either the dog's submission or any of our own guilt-based behavior lies beyond our abilities, professional help may be in order. And obviously, if we feel that recognizing guilt in our dog and/or ourselves is beneficial to our relationship, attempting to change our own or our dog's guilt-based behavior is bound to fail.

Finally, we can consider getting rid of the guilt. We can refuse to see it in the dog and not acknowledge it as a workable emotion from which to generate beneficial body language. We can learn to ignore it in ourselves or replace it with something more productive. Easier said than done, to be sure; but a dedicated owner can do it. Let's remove the guilt form Lou's original interaction with Merlin and see how differently the chewed-glove incident unfolds:

One evening Lou, in such a hurry to get to the celebration banquet, leaves his prized glove where Merlin can reach it. When Lou returns home late, he finds the shredded mitt in the middle of the floor. Enraged, he bellows and lunges at the dog, who freezes, lays back his ears, and tucks his tail tightly against his abdomen. Seeing the dog's submissive response, Lou recognizes his dog's fearful response to his own anger and pauses, then sinks down on the couch. Picking up the remains of the glove, he sees his own tears glistening on the shredded leather, while Merlin, sensing the change in his owner's demeanor, slowly creeps toward his master and gently plops his big black head on Lou's knees.

That night it takes Lou a while to fall asleep. "If only I'd put the glove away" alternates with "If only I'd taught Merlin not to chew." Eventually, however, he falls asleep. When he awakes the next morning, the loss of the treasured glove still saddens him, but along with his regret he also feels a sincere determination to teach his dog not to chew.

In this scenario we see how recognizing the submissive canine response enables Lou to put the problem in its proper perspective. Because the problem is the chewing—not the guilt—Lou honestly deals with what's bothering him: the loss of a treasured object and the need to train his dog.

When Lou enters the living room and sees Merlin and the glove, he has two problems to confront *immediately:* the chewing, which it's too late to deal with because the dog finished off the mitt hours ago, and Merlin's fear, evidenced by the dog's submissive response *now.* That a valued object has been destroyed and that he scares his own dog with his angry outbursts both justifiably sadden Lou. Better to manifest his genuine sadness than mask it as righteous anger directed against his dog. Only in such a way can Lou free himself from one problem so that he can deal with the other. Having mourned the loss of his beloved glove and the fact that his interactions with his dog in the past have led Merlin to fear him, Lou can now objectively evaluate Merlin's and his own behavior and make lasting changes.

Although getting rid of the guilt offers the most positive long-range solution, getting rid of the dog makes sense in some (few) instances. If the guilt the owner believes the dog exhibits or the owner's own guilt arising from the relationship is unbearable, and accepting, changing, or eliminating the guilt is impossible, the relationship is doomed. Remember our devout pacifists in chapter two? When one of their dogs killed his littermate, two guilt-based responses poisoned the owners' relationship with the survivor:

- They couldn't believe the survivor could kill his "own brother" without exhibiting any signs of guilt.
- They couldn't accept, alter, or ignore their belief that they were indirectly guilty of killing their own dog.

Our dogs frequently serve as constant reminders of our human shortcomings as well as our strengths. When they reflect what we consider our better selves, our self-image and our relationship with our pets blossom; when they make us see the worst in ourselves, or when we can only see the worst in them, our relationship withers to the point where termination offers the most humane solution for both human and canine.

THE BONDED VIEW

Recall how we painted a better scenario for Lou Rutherford's alternate response to Merlin and the chewed mitt, the scene in which he manifests only his sorrow over the loss of the mitt because he realizes it's too late to discipline the dog meaningfully. In this ideal bonded response, Lou concentrates on the only two pertinent issues: the glove and his relationship with his dog. Finding both less than what he wants, he responds with genuine grief, then determination to rectify the underlying problems.

The bonded view has not yet come up in our discussion of guilt because bonded owners don't perceive guilt as a workable emotion in their relationship with their pets. If their dogs exhibit what they consider to be negative behavior, they respond to the behavior. They don't waste time blaming themselves or see guilt in their pets; they concentrate their efforts on solving the problem.

This doesn't mean that bonded owners don't get angry when their pet misbehaves. Nor does it mean that bonded owners never overreact to their pet's misdeeds, lash out, and feel sorry later. But note the subtle and critical difference: If they can't accept the behavior, bonded owners feel anger and/or sorrow and manifest the accompanying body language, not guilt.

Most of us respond half-bondedly. We openly grieve or feel anger over the results of our dog's behavior; however, if we feel responsible for what happened in any way, we seek some sort of retribution to ease those feelings. We delete all those steps in the chain of events that reflect on our lack of consistent positive interaction with our dogs and concentrate only on what the dog did to hurt or inconvenience us. However, if we value our pets, as most of us do, we can't escape our responsibility for any behavior our dogs display.

Invariably we must come to grips with the fact that owner

and dog work as a team, reflecting each other's frailties and strengths. When Emily yells at Bit O'Honey for jumping up on her neighbor because she knows her friend dislikes that behavior and Bit O'Honey cringes and tucks his tail, a bonded Emily doesn't read guilt in her dog's body language. She knows he's confused and responding submissively because she lets him jump on her and even encourages it. The only way she can remove the ambivalence from their relationship is either to allow Bit to jump on everybody or to train him to jump on no one or only on a select few. Regardless of her choice, only Emily can make and implement it. Unless she deals directly with the basic behavior rather than its results on a hit-or-miss basis, she can't build a stable bonded relationship with Bit O'Honey.

We can also think of guilt as a red herring, steering us away from a bonded relationship. Just as these smelly fish were used to lead hounds away from their quarry, so guilt causes us to expend vast amounts of time and energy without making the least headway toward our goals: solving problem behavior or mending shaky relationships. However, while guilt itself serves no useful purpose in solving problems, often it does provide an instructive marker, telling us we're not dealing with either the real problem or the real emotions. If you're one of the many owners who perceive guilt in your dog's behavior or your responses to it, don't berate yourself. Simply recognize the body-language expressions and associated feelings, assigning them more manageable and productive meanings. It might be difficult at first, but keep reminding yourself that submissive body language and guilty feelings evolve from your relationship with your dog on the most intimate one-to-one level—not from anything either of you does or doesn't do. Armed with this recognition, you can convert a nightmare emotion into one that can make your dream of an ideal relationship come true.

In the next three chapters we're going to take a more intimate look at three important behavioral states—submission, dominance, and fear—each of which carries strong emotional charges for most dog owners. While these states serve stabilizing and protective functions for the wild dog, in domestic situations our emotional interpretations often transform them into highly destabilizing and even destructive forces.

We'll begin by discussing submission and its related emotional states, dependence and devotion. Do you love the way Fumbles adoringly sticks to you like glue? Do you hate it when he cowers? Would it surprise you to learn that the behaviorists don't differentiate such seemingly different displays?

5
SUBMISSION, DEPENDENCE, AND DEVOTION: ALL OR NOTHING?

EVEN THOUGH SEVERAL PEOPLE advised Pamela and Terry Pederson not to get littermates, the first-time owners couldn't resist Rocky and Lily, the two remaining pups in a litter of eight springer spaniels. Rocky is so bouncy and full of energy, they can already see themselves frolicking with him in the park. But how can they possibly leave poor Lily huddled in the corner all alone?

During their first year with their pets, the Pedersons receive a crash course in canine behavior. While both pups have their problems, shy, cowering Lily initially complicates her owners' lives the most. Every time Terry raises his voice, the pup flattens herself against the floor. The few times he disciplines her, she rolls over on her back and urinates submissively, which only angers her owners even more. Although in the beginning Pamela is quite touched by the way Lily follows her everywhere, she eventually grows tired of finding the pup constantly underfoot.

"Go and play with your brother," she scolds, shooing Lily outdoors where Rocky is busily chasing squirrels around the yard. Lily fixes her mistress with her soulful eyes, presses herself tightly against the door, and whines softly.

"What a wimp!" Terry fumes in disgust. "How can two dogs from the same litter be so different?"

We've already discussed the behavioral state of submission in general terms, but now we're going to explore some of the emotions and meanings we often assign to these displays in more detail, and for a very special reason. Because our pets almost invariably display submission around us or others, we must learn to recognize these canine behaviors as well as our human emotional responses to them. We already know that the dog's social nature predisposes it to act either dominantly or submissively in certain situations. However, whereas pack rank tends to be fairly stable and predictable in wild dogs, the rank of household pets may fluctuate wildly. In one situation a dog might assume immediate control; in another, it might be incapable of moving a muscle without our reassurances. Sometimes we praise and other times we punish identical displays. Given such inconsistency, we can't possibly hope for an ideal relationship with our dogs unless we learn to recognize the body-language cues, reinforcing those that benefit and minimizing or eliminating those we think might harm the relationship.

Bearing in mind that our dogs readily incorporate people into their pack structures, let's add a new dimension to our dependent- and independent-owner orientations. While most of us theoretically accept that humans totally dominate canines if for no other reason than that we're responsible for our pets and not vice versa, we don't always act as if we really believe this to be true. While on the one hand we expect to provide food, shelter, and veterinary care, we don't necessarily expect to run interference for our dogs in dog fights; and sometimes we even expect our pets to run interference for us.

In light of such conflicting expectations, we can see how an owner's reaction to a dog's submissive display depends on whether or not he or she consciously or subconsciously relates in either a dominant or a submissive fashion to the animal and event. If we expect the dog to be submissive (that is, we accept the dominant or leadership role), we view a submissive display entirely differently than if we expect the dog to take charge. In the latter situation, we wind up with two followers and no leaders, exposing ourselves and our pets to all the negative emotions inherent in any situation where the blind lead the blind.

In general, dependently oriented owners are also more submissively oriented. While on the surface this might seem para-doxical because such individuals view their pets as furry hu-manoids, the two naturally share their dependency. To be sure, dependent owners may fuss over their dog's most minor real or imagined needs, but they also tend to depend on their pet to validate their own sense of well-being. Such an approach cre-ates inconsistent and contradictory demands, which even the most stable pets can't possibly fulfill. On the one hand, these owners want a furry, dependent baby they can cuddle and care for; on the other, they expect their coddled pets to support and protect them. When Lily jumps back in surprise as a squirrel dashes across her path, Pamela laughs, picks up the pup, and coos, "Silly Lily, do you think I'd let that mean old squirrel hurt my little girl?" When Lily signals her fear of the paperboy by barking and growling anxiously, Pamela congratulates the pup for being so courageous: "What a brave little trooper you are!"

In sharp contrast, independent owners, who view their pets as a separate species that must be controlled, usually respond dominantly to their dog's submission. While in theory this should establish a stable relationship closely resembling wild-dog-pack dynamics, that doesn't always happen. Even indepen-dent owners, who prefer to call the shots under most circum-stances, invariably define certain situations as ones in which the dog should take command. And, like most owners, they expect the dog intuitively to recognize those special situations.

As we discuss the body-language expressions of submission, dependence, and devotion, always bear in mind the relative na-ture of these behavioral states. The singular dog acts submissive, dependent, or devoted only in relationship to people and events. Whether we perceive the display as acceptable or not depends on whether we can accept the relative dominance the dog's submission confers on us.

THE CANINE BODY-LANGUAGE SIGNALS

We've already discussed the body language behaviorists as-sociate with submission, but let's quickly review the most com-mon signals:

- Ears flattened against head.
- Eyes averted.
- Tail down.
- Rolling over and exposing the abdomen.
- Dribbling of urine.

When our dogs display these signs toward other dogs, we may or may not pay much attention. We may simply note that dogs naturally greet each other this way, or we may assign a whole host of emotional meanings to the display. For example, the belly-up submissive posture derives from the newborn pup's first experiences. When a pup finishes nursing, the bitch rolls it over and licks it; this lapping of the area around the rectum, penis, or vagina stimulates the pup to urinate and defecate. The bitch then ingests the excretions as part of a process that effectively and efficiently keeps the nesting area clean, helps control parasites, and prevents the spread of disease. In addition to serving a valuable physiological function, this maternal behavior lays the foundation for one of the most basic pack behaviors: From the very first day the pups respond to authority (bigger, older, stronger, more experienced Mom) by displaying the exposed-abdomen position. Consequently dogs don't need to learn the proper way to greet dominant animals. To them it's as natural as their most basic bodily functions. As they grow up, the behavior not only helps maintain their physiological health, it also helps protect them from attack or abuse by other pack members.

While no one would deny wild dogs the beauty or efficiency of such a system, many people can't accept it in their household pets. Owen Chancellor purchases three-month-old Bayberry, a relatively submissive and very intelligent Kerry Blue Terrier from a respected breeder. Taking advantage of both her intelligence and submissiveness, Owen quickly teaches his new pup a very special trick—to roll over in response to a subtle hand signal. Anxious to show off his smart pup, he takes Bayberry to the Chancellor family reunion.

Trotting the immaculately groomed terrier up to a group of youngsters, he tells them that when Bayberry meets good little boys and girls, she always rolls over so that they can scratch her tummy. As the children watch the dog expectantly, Owen touches his index finger to his nose, and Bayberry drops and rolls. The children, of course, squeal in delight.

Knowing that his well-trained pet won't wander far, Owen lets Bayberry roam freely among the throngs of relatives. First the Kerry Blue approaches Uncle Oscar, the asthmatic and allergic anticanine curmudgeon of the clan. As Bayberry approaches, the old man feels a sneeze coming on and raises his hand to abort it. Bayberry dutifully rolls over and exposes her abdomen.

"*Aaachooo!* Cowardly cur, afraid of a sneeze!" snorts Oscar, aiming a swift kick at the prone dog.

Bewildered, Bayberry scurries away toward a group preparing the picnic supper. As Cicely Chancellor lifts the lid from a steaming pot, she uses her free hand to brush perspiration from the tip of her nose. Once again the ever-vigilant Kerry Blue responds to the subtle cue.

"Oh, you poor dog," Cicely exclaims. "Someone must have beaten you. Here, let me get you a hot dog."

Still munching her treat, Bayberry seeks out her owner, who has been playing cards on the porch. "Want to see a really smart dog?" Owen asks his pinochle pals. Again he flashes the signal. Again Bayberry goes belly-up—this time to an appreciative audience awed by the dog's performance.

In such ways a simple body-language expression genetically and environmentally tempered to protect the wild dog becomes enmeshed in a web of complex human emotions:

- Owen bursts with pride when he brags about his dog's intelligence.
- The children squeal delightedly at Bayberry's great love.
- Uncle Oscar is infuriated by Bayberry's cowardice.
- Cousin Cecily sympathizes with the dog's past suffering and vulnerability.
- The cardplayers admire her superior intelligence and obedience.

The canine body language we associate with dependency can be even more complex, if for no other reason than that it often falls within the nebulous realm of "looks," which include those we often associate with great canine need: beseeching glances or stares, trembling, whining, hesitation. In addition, we sometimes define dependent displays as those behaviors we believe a dog can and should perform alone but won't.

For example, some dogs refuse to eat or drink in the owner's absence, even though both food and water lie close at hand. Others refuse to relieve themselves unless the owner accompanies them; still others won't sleep anywhere but in a particular spot, with a particular toy, or in their owner's presence. Surely we've all seen or perhaps even been among the scores of shivering owners waiting in the middle of a totally secured yard for the dog to relieve itself. There's no physiological reason why we shouldn't be sitting in the warm kitchen instead of standing in the freezing rain except that everyone knows Peggy Sue won't go when she's alone: She'll hold it until she comes back inside, then she'll go on the floor.

The key to many definitions of dependent body language rests with a belief that a dog demands our presence at, or assistance in, a behavior we don't wish to share. Compare this to devotion in which our pets display needs we *want* to fulfill. For every example we use to describe dependent behavior we can find owners who define the identical behavior as proof of their pet's devotion. Therefore, although the differences between dependent and devoted behaviors may escape the casual observer, they're quite apparent to individual owners.

IT ALL DEPENDS ON YOUR POINT OF VIEW

The dependent owner of a submissive pet experiences a schizophrenic relationship marked by the highest highs and the lowest lows. Bachelor Richard Wilcox inherits Muffin, his mother's five-year-old poodle, and at first the dog is a source of great comfort. Every time Muffin runs trembling to his side when the doorbell rings, or scurries under the bed at the sight of a visitor, Richard rushes to console her. He even brags to friends that she refuses to eat until he prepares his own meal and sits down at the table. Because she's a small dog, he doesn't object to her insistence upon sleeping right next to him. After all, didn't he consider that one of Muffin's most devoted acts when she lived with his mother? True, he doesn't find the sound of her constantly chewing her squeaky toy as cute as his mother did, but that's a minor annoyance. Or it would be minor if it didn't increasingly grate on his nerves. Before long the habit becomes like a tiny cut that won't heal; and every time he acknowledges his feelings about it, he packs the salt of guilt into the wound.

Obviously Flora Wilcox didn't encourage the behavior with the idea it might someday torment her son, but why did she do it? Let's take a look at how Muffin's displays evolved, and the meanings and emotions her past and current owners assigned them.

Because she lived alone, Flora wanted a pet she could fuss over. Whenever she heard a loud noise, she rushed to the trembling Muffin, cuddling her and cooing softly, "There, there, baby Muffin, Mommy's here. Nothing's going to hurt my little woofkins." She spent hours brushing and combing Muffin's furry pom-poms; when Muffin went to the groomer's, Flora had her own hair done in the beauty parlor next door, "so Muffin won't be afraid." Because Flora prided herself on her independence in a community of young up-and-coming professionals who tended to treat anyone her age as dependent, she did everything to reinforce Muffin's dependency: "Maybe Richard doesn't need me anymore and thinks I have to be looked after, but that's not true. My little Muffin needs me, and I'm perfectly capable of taking care of both of us," she'd confide to Muffin as she placed the only brand of dog food she believed good enough for Muffin in the poodle's crystal bowl on a handwoven placemat in an honored spot next to her chair.

The resultant relationship benefited both dog and owner. Flora took great pride in her ability to care for her pet and saw many of Muffin's submissive displays as evidence that the poodle needed her care; this also reinforced her definition of herself as independent. By defining other submissive behaviors as indicative of Muffin's devotion, Flora had visible proof of her dog's love. During the five years they were together, the relationship worked because their needs fit together as snugly as two pieces of a jigsaw puzzle.

However, after six months with the dog, Richard considers calling it quits with the relationship. "I'm sick to death of that dog following me everywhere I go. I used to think it was funny, but dang it, I'm *not* her slave. She can be such a demanding little bitch. And she's afraid of her own shadow." By the end of a year of uneasy compromise, Richard has generated a whirlwind of negative emotions around what his mother considered perfectly normal behavior. How did those first guilty twinges in response to Muffin's squeaky toy evolve into such a violent storm?

Muffin's story illustrates the problems that can occur when one owner supports and even encourages submissive and dependent behaviors another can't or won't. Richard is a twenty-seven-year-old lawyer; he doesn't want a wolf, but he doesn't want a chicken either. Although he can initially tolerate Muffin's behavior by viewing it and her as a legacy, that doesn't work for long. Remember the origin of the "white elephant"? In some cultures white elephants were considered sacred, and whoever owned one was obliged to give it the best possible care. Those who wanted to get even with their enemies "honored" them with such a sacred gift, hoping the burden of its care would ruin the owner both financially and emotionally. In such a way, a gift can become a curse and the giver someone for whom your feelings must eventually deteriorate.

If Richard redefines Muffin's dependent and devoted displays as demanding, manipulative, and cowardly, one of his strongest ties with his mother threatens to strangle him. During one memorable month he spent at least five hours trying to coax Muffin out from under the bed to greet former friends of his mother's who wanted to see the dog. While his visitors found the sight of one of the area's most rapidly rising legal stars wriggling under the king-sized bed and flailing at the terrified dog with a broom quite funny, Richard was not amused. Nor did he laugh when he had to:

- shorten two business trips because Muffin refused to eat or drink at the boarding kennel.

- cancel his vacation because she'd starve herself without him.
- miss a long-awaited concert when he had to track down Muffin's special blanket, which had somehow gotten mixed up with the dirty laundry.

And he takes no comfort in the fact that his mother would have found such inconveniences endearing and desirable. Richard wants a submissive pet only in the sense that he wants a dog that obeys. He has come to loathe Muffin's displays, seeing them not only as excessively submissive to the point of being cowardly, but as a nefarious plot instigated by the poodle to dominate him.

WHO'S IN CHARGE HERE?

While Richard may be awash in anger, guilt, and frustration, accusing his dog of being a pint-sized canine Machiavelli, the real problem is that he hasn't established leadership of this pack. Without clear lines of dominance or submission, no consistent relationship can develop.

Reflect a moment on the eccentric but nonetheless stable relationship Flora Wilcox enjoyed with Muffin: She wanted her dog to depend on her and she reinforced the behaviors necessary to maintain this belief. When Richard inherits Muffin, he also inherits Muffin's orientation toward humans and events; however, he doesn't want or need a dependent dog to maintain his image of himself as independent. Therefore he can't respond to Muffin the same way his mother did, nor does he want to submit to the poodle's demands. Because he sees these as his only two options, he chooses to do neither consistently and instead relates to the dog on the basis of his moods. Consequently he may console Muffin when she cowers on Monday, ignore the same display on Tuesday, and punish her for it on Wednesday. From Muffin's point of view, this means that Richard's submissive on Monday, no one's in charge on Tuesday, and Richard's dominant on Wednesday. Imagine if such capricious swings of authority occurred in your business or home: If you wanted simply to accomplish your assigned role to the best of your ability, such leadership gymnastics would leave you feeling frustrated and confused.

"I just want to do what's best for Muffin," groans exasperated Richard. That's a very good attitude with which to begin solving dependency-based problems, but we must take it one step further: Muffin is a *dog*, albeit a bright, charming one, but a dog nonetheless. Therefore Richard can only guess what's best for her, and he will most likely base his guesses on prejudiced personal experience, intuition, or even scientific studies. While such guesses and opinions have merit, the survival and health of the relationship depends equally on Muffin's *and* Richard's happiness. And because we can at best only guess what Muffin wants, doesn't it make more sense for Richard to decide what *he* wants?

"I certainly don't want Muffin to dictate my moods and behavior," states Richard firmly. "On the other hand, I can't bear to see a dog cower every time the owner looks at it—I could never dominate a dog that way." We can see how, given his limiting definitions of the options available to him, Richard would resist embracing either one.

However, let's add a third option: dominion. Within the literature and lore of all religions, humankind is given dominion, not dominance, over the creatures of the earth. If we accept such a responsibility, our proper relationships with our pets aren't those of master and slave—regardless which species assumes which role—but rather reflect our benevolent sovereignty: We should do our best to provide guidance and accept responsibility for the actions of our pets. Dominion over our pets means we *want* to care for, guide, and accept responsibility for them and their actions, and that we recognize them as being both different and unique. Dominion demands creativity, self-confidence, respect for the other, and a belief that the relationship is mutually beneficial and worth preserving.

LILY-LIVERED LILY

Before we help Richard establish dominion over Muffin, we need to examine the invariably negative associations many people make between submissive displays and cowardice. It's one thing when Uncle Oscar takes a swipe at Bayberry once a year when the Kerry Blue exposes her abdomen, but it's something else entirely when Lily's submissive displays turn the Pederson household into a battlefield.

In the Pedersons' case we encounter multiple relationships at work, the most stable and unemotional of which occurs between Lily and her littermate, Rocky. Rocky is dominant, Lily is submissive, and the two get along quite well. However, as all owners know, it's difficult to keep human beliefs and prejudices from sneaking into our relationships with our pets. The Pedersons initially took Lily because she was so "shy," "vulnerable," and "delicate," beliefs they revealed when they named her. Up until the pups were four months old and Pamela and Terry decide to institute "serious training," they laugh at Lily's fearful and skittish behavior. "What a goofy dog!" chuckles Terry, letting the pup huddle beside him. "Such a delicate little flower," croons Pamela as she rocks the trembling pup during a thunderstorm.

Most inexperienced owners expect their pups to outgrow submissive, dependent behaviors. However, no owner can expect a pup to shed any behavior the owner has deliberately or inadvertently reinforced. If the Pedersons condone Lily's behavior now but won't accept it from an adult dog, they must realize they're creating their own problems by tolerating the display, especially during that critical learning period when Lily is between two and four months of age.

When, as part of the later serious training, Pamela yells at Lily for cowering instead of cuddling her, the pup is baffled. When Terry swats her, Lily's fear overcomes her and she rolls over and urinates submissively. This infuriates Terry, who smacks her again and storms out of the house.

"How can you be so stupid?" cries Pamela, crouching on the floor beside the trembling pup. Without realizing what she's doing, she absently begins stroking Lily's tummy as she flips through the training book to see what she and Terry are doing wrong.

Who's in charge in this situation? Initially Lily occupied the most submissive position in the household pack: Rocky, Terry, and Pamela were all more dominant and consequently rewarded Lily's submissive displays. When the training begins, the Pedersons decide they want Lily to lose her shyness and behave more like Rocky. Right away Lily's trapped in a dilemma: She and Rocky get along well precisely because she *doesn't* act like him. Furthermore, her submissive orientation results from sixteen weeks of highly complex interactions, first with her mother and all seven littermates, and then with Rocky and the Pedersons. Any attempt to switch off this behavior suddenly in order to make her equal to her brother and dominate events her owners previously allowed her to react to submissively creates tremendous stress on the pup.

To further complicate matters, the Pedersons get caught in an all-too-familiar, nonproductive tug-of-war with their submissive pet. The angrier they become, the more threatening they appear to Lily. The more threatening they appear, the more submissive the pup's displays. The more submissive the pup's displays, the angrier the owners become. Eventually owners either physically punish the dog until guilt aborts the cycle or they alienate it or terminate the relationship to avoid such painful consequences.

"We could never beat Lily, no matter how cowardly she acts," declares Pamela with conviction. "She's just a baby."

"Maybe not now," says Terry slowly. "But I'll be darned if I'll put up with a paranoid dog. We got them—both of them—to take care of you when I'm on the road. That's the least Lily can do. I don't see why she can't act like Rocky—it's in her genes, for goodness' sake!"

FINDING THE SOLUTIONS

As with most problems we encounter in our daily lives, an attitude of confidence plays a critical role in solving problems that undermine our most basic relationships with our dogs. If guilt starts us off on the wrong foot, renewed confidence can put us firmly back on the right path. But how can we build our confidence when a relationship has been shattered, we're convinced our dog is incorrigible, and we must guiltily acknowledge that, at least some of the time, we feel the same way ourselves?

The answer lies in having confidence in what you want. You must determine what you really want from the relationship with your pet. Obviously Flora Wilcox and Richard did not want the same relationship with Muffin. Does Terry want Lily to relate to him the same way she does to Pamela or Rocky? Does Pamela expect Lily to relate to her differently from Rocky or Terry?

Once we know what kind of relationship we want, then we can complete the research necessary to learn whether we can achieve it. For example, Richard wants Muffin to like and obey him, but he doesn't want her slavish devotion or extreme dependency. He feels confident that some middle ground would benefit them both. Consequently, when he finally calls a local trainer for advice, he's able to articulate his needs clearly. He feels no need to apologize for Muffin's behavior, his reponses, or his choice to make necessary changes to achieve his desired results.

Because of his open and honest approach, Richard quickly wins the support of the trainer, who helps him set up a program of daily confidence-building exercises. "Bear in mind it took Muffin five years to develop her present behavior, so don't expect her to change in a week," warns the trainer. Both trainer and owner also agree that Muffin's dependence and fear of strangers could easily evolve into more fear rather than confidence if anyone but Richard initiated the training.

Muffin's program aims to make her less dependent on her owner and more accepting of other people and events. By teaching her to respond to some simple commands, Richard accustoms her not only to being obedient but also to performing a task and succeeding at it. True, Flora Wilcox never felt Muffin had any behavioral problems, but she never asked the poodle to do anything either. Consequently, most of Muffin's contribution to that relationship centered around those looks and body-language expressions the dog used to communicate her needs. Now Richard hopes, via confidence-building training, to replace his mother's "What can I do for you?" orientation toward Muffin with a more equitable "What can we do for each other?"

Although it takes almost six months, Richard teaches Muffin to come, sit, and stay on command, not only in the security of their home but also in a nearby park and small shopping center. During this period Richard also asks a neighbor to stop by every day, even if only for a few minutes. At first Muffin runs for cover under the bed, but when Richard doesn't follow and the same person appears every day, her curiosity eventually propels her from her hiding place. In a few weeks she retreats only to the bedroom doorway and hardly trembles at all.

Once Richard gains confidence in Muffin's responses to him, he invites his neighbor to appear shortly after the evening training session has commenced. At a prearranged time Richard commands Muffin to sit and stay as the neighbor rings the doorbell and enters the house; Richard holds the poodle's gaze and repeats the command. Eventually Muffin performs flawlessly in the neighbor's presence and soon learns to respond obediently to the neighbor as well. By the end of her first year with Richard, Muffin responds consistently to four people and tolerates his absence and occasional kenneling with minimal stress.

What about all those obnoxious dependent behaviors that drove Richard crazy—shouldn't he train her not to exhibit those? Because Richard recognizes that these behaviors spring from Muffin's lack of confidence, he concentrates his efforts on eliminating their cause rather than their numerous negative effects. To be sure, it's possible to teach a dog not to cringe every time the doorbell rings or not to follow the owner everywhere—but it's not easy. For one thing, we must start with a dog that's submissive and lacks confidence and punish it for body-language expressions that reflect those mental states. While we may be able to extinguish certain behaviors with sufficient punishment, we do nothing to enhance the dog's independence and may even undermine its low level of confidence even more. It's far better to concentrate on building the submissive animal's confidence via consistent training programs that enable both owner and pet to succeed, because this route alleviates annoying dependent behaviors almost miraculously.

Lily and the Pedersons spawn a different sort of confidence problem. Whereas Richard saw Muffin's shortcomings as a result of her dependency, the Pedersons view Lily's behavior as an expression of her cowardly nature. And although many people might agree that one dog's dependent behavior is a sign of devotion in another, only a fool would dare challenge any owner's interpretation of his or her pet's cowardly displays.

Why? Because cowardice is so purely subjective and such a highly charged negative emotional state. However, behaviorists don't even recognize it as a valid canine condition. Rather, they classify these so-called cowardly displays as either submissive, dominant, or fear-based; only unknowledgeable emotional humans define them as cowardly. The Pedersons face a difficult problem because they committed a common error, not only equating submission with cowardice but also relating dominance to bravery. Therefore they initially define the solution to their problem as making Lily act like her dominant littermate; but if Lily does that, the stable relationship she now enjoys with Rocky is bound to suffer.

Terry and Pamela need to progress beyond their human emotions and see the problem objectively. If they truly understood pack dynamics, they would never want their pups to be "equal"; that would guarantee continual antagonism between the dogs. What they really want is a more confident pet, not a dominant one.

Here again knowledge and basic training work wonders as confidence builders. Once Terry and Pamela realize that Lily's submissive temperament is different—not wrong, abnormal, or cowardly—and critical to her stable relationship with Rocky, they begin to see her in a completely new light. Simply recognizing certain displays as normal for Lily enables them to ignore them. If her owners reduce their yelling and punishment, Lily will naturally grow more confident and outgoing. The Pedersons also discover that Lily prefers slower, softer training approaches rather than the fast-paced ones Rocky enjoys: Just because the pups are littermates doesn't mean they *have* to be trained the same way.

LESS THAN LEAST

Perhaps the most troublesome relationship occurs when a submissive dog is matched with a submissive owner. In such situations, the owners don't want their dogs to act like dominant dogs; they want them to act like dominant *humans*. From within the ranks of such relationships emerge the most ill-prepared, ill-motivated "protective" canines. The scenario typically includes a single adult—usually female with or without young children—and a dog obtained "for protection." The dog is usually big and even fierce-looking—Doberman, shepherd, or husky type—and chosen because he or she barks savagely. Although animals of both sexes can exhibit such behavior, the submissive owner usually picks a male because "everyone knows males are better protectors."

Let's watch what happens when Cathy McCaffrey moves to Chicago following her divorce. She "rescues" a timid shepherd mix from a shelter and quickly wins the dog's devotion with her lavish attention and genuine affection. Whenever the skittish Czar barks at the slightest noise, Cathy praises the behavior but ignores the dog's trembling and worried expression. She doesn't realize that Czar isn't a dominant dog, let alone a dominant human; nor does she recognize that his frequent belly-up postures indicate something besides his desire to have his tummy rubbed.

Czar is submissive by nature and by experience. Were he a wild dog, the pack leader would deal with any perceived threats, or at least tell him what to do. When he barks in fear or panic, his owner rewards him; that connection defies Czar's canine logic: Why does it please her when he's afraid? Before long, confusion accompanies his fear, and his natural submission takes a bizarre twist.

By the time Czar reaches one year of age, three emotions dominate his life—fear, confusion, and devotion—all arising

from his owner's interpretations and reinforcements of his submissive displays. Because Cathy herself feels insecure and fearful, she wants a dog for protection. However, her limited
knowledge of normal submissive canine behavior and her highly
prejudiced and erroneous views of both protection and devotion
inadvertently set the scene for disaster.

Those who share Cathy's experience always express shock
and surprise: "I don't know what happened. One minute my
dog was beside me, the next he'd clamped his jaws on to the
meter reader's arm." Cathy was walking Czar toward her home
one quiet summer evening when a friend wearing a garish hat
jumped from behind a bush to surprise her. Czar lunged and
connected, and when Cathy tried to pull him off, she was badly
bitten herself.

Dramatic? Perhaps, but hardly uncommon. Unfortunately
for the many unknowledgeable owners of such animals and
their unwary friends and neighbors, such indicents happen all
the time.

The proper philosophy of protective- or attack-dog ownership can be summed up very simply: If you have the knowledge,
confidence, and skill to properly handle such a dog, you probably don't need one. And the opposite also holds true: The more
fearful and less confident you are, the less qualified you are to
handle a protective dog. In other words, the more you think you
need one, the more you should avoid owning one.

One highly reputable trainer responds to inquiries for attack-dog training by urging those people to discuss the consequences with their lawyers and insurance agents. That
approach reflects common sense: It was far, far more likely that
Cathy and Czar would encounter nonthreatening people and
friends than a mugger or worse. However, as every veterinarian
who has had to euthanize such dogs knows and as every emer-
gency-room physician who has stitched up meter readers, letter
carriers, and children can attest, frightened people seldom follow the most logical or sensible path.

To be sure, protective canines also arise from the dominant
ranks, and they create their own problems, as we shall see in the
next chapter. However, the submissive dog forced into protec-

tive dominance by a fearful owner can become so unstable as to be extremely unpredictable. Imagine yourself as a 110-pound pastry chef who's suddenly yanked from the bakery and told to protect a 250-pound linebacker. When you jump every time you hear a noise, your protective responsibility shouts, "Wow, that's great! Keep up the good work!" Despite your uneasiness with the assignment, the linebacker is a nice guy you want to please, and he pays you well. But in the back of your mind the conflict grows unbearable: What am I doing here? Why can't he take care of himself? Shouldn't he be taking care of me? Why did he get mad at me when I threatened his girlfriend?

To the submissive dog, regardless of sex or breed, all humans look superior. If these dogs respond protectively, they probably do so out of fear rather than any great love or devotion. Even if we believe the response arises from the latter, that belief in no way gurantees it will be the proper one or aimed at the proper target(s).

The only way Cathy can guarantee Czar's response is specifically to teach it to him. In order to do so, though, she must first come to grips with her own fears. Chances are she can't because chances are what she's afraid of is the unknown.But Czar's perception of that unknown differs radically from her own; his vision is ten times more sensitive to motion, his range and intensity of hearing much greater, his ability to recognize scent so incredible we can't even comprehend it. So when Cathy reinforces his fears, she participates in the creation of a canine "unknown" that could easily terrify her dog much, much more than any human "unknown" might frighten Cathy.

Furthermore, if she succeeds in teaching him proper protective responses and he gains the necessary confidence to ensure a consistent display, he'll probably give up many of the fear-based signs she associates with his protective displays. Is Cathy willing to have her dog relinquish those body-language expressions? Or do her own fears plague her so much that she'd rather perpetuate her dog's body language and the erroneous meanings she attaches to it? Far, far too often our own fears and definitions determine how our pets manifest their natural submissive orientations.

In such cases, the owner must be willing to change before attempting to change the dog, and such a change may require professional counseling. If that isn't possible, the truly compassionate owner should terminate the relationship; otherwise dog and owner will simply feed off each other's fears until an unfortunate incident leads to the destruction of the relationship anyway.

Let's consider another example. Scott Vickers simply can't accept that his belly-up urinating male golden retriever is anything but a coward, even though he knows what the behaviorists say and he finds their arguments sound. As the most submissive pup in a litter of ten, his dog Atilla responds submissively to man and beast alike. If Scott raises his voice, a puddle of urine appears beneath the trembling pup. As if to apologize, Atilla then tentatively wags his tail, splattering urine everywhere. Every time Scott and Atilla encounter one of the pup's littermates in their small community, inevitable comparisons arise, and Atilla's always found lacking: What a poor choice of dog for the area's best alpine skier and wind surfer!

When an owner perceives his or her image so threatened by perplexing and unwanted canine behavior, termination again becomes a viable option. Even though Scott thinks he could have worked out a more equitable relationship with Atilla given enough time, he resists an impulse to give the pup another chance. The young couple who take Atilla are thrilled with the idea of working with him, no matter how much time and effort it may take. Scott feels no regret as he watches the now-confident Atilla interact with his owners and others a year later. His choice clearly benefited him and Atilla. Sure, Scott might have created a great relationship with Atilla; but chances are it would have become a lot worse before it got any better. By making the choice to terminate the relationship when Atilla is still a pup, Scott gave the dog and his new owners an opportunity to create change without trauma. Those who would fault Scott should remember that it often takes just as much confidence to terminate a bad relationship with a pet as it does to improve one—and sometimes even more.

THE BONDED VIEW OF SUBMISSION

My husband tells the story of a particularly boisterous fraternity party back in the days when propriety placed curfews on young ladies. This specific evening, partygoers lost track of the time, and when the police arrived in response to a neighbor's complaint, panic set in. One mortified coed hid herself in a refrigerator with the door slightly ajar. When an officer opened the door and demanded to know what she was doing in there, the distressed student stammered, "Everyone's gotta be somewhere."

When it comes to pack position, that's a very valuable point for all dog owners, but especially those with submissive pets, to keep in mind. If our dog behaves submissively, we should not; otherwise no one's in charge. Nor are we doing our dog any favors by reinforcing its fears to make ourselves look dominant or feel more secure. When Richard Wilcox takes the time to recognize what he wants from his relationship with Muffin and initiates a program of changes necessary to form a strong bond with his dog, he accepts the responsibility for his dominion over his pet. When the Pedersons subordinate their highly emotional, erroneous, and negative evaluations of Lily's submission, they simultaneously acknowledge their role in Lily's view of her position within the Pederson pack. In both cases the owners change their own orientations toward submissive and dependent behavior in their dogs and then create lasting and stabilizing changes in their relationships.

In a far different but nonetheless impressive way Cathy McCaffrey and Scott Vickers also recognize the value of a bonded relationship. Cathy was willing to undergo several years of often grueling counseling to deal with her fears, as well as the time and expense necessary to initiate and maintain a quality professional training program for Czar. Had she not valued a bond based on dominion rather than mutual fear and dependency, she would never have gone to all this effort.

And while there may be those who would prefer to castigate the Scott Vickers of the world as irresponsible villains, the fact that he chose to terminate his relationship with Atilla tells us he recognized that what was going on between him and his dog wasn't acceptable. He knew things could and should be a lot better and he had the confidence to accept that he was unwilling to change. By so doing, he extricated himself and his dog from a dead-end, unfulfilling relationship—and he freed his pup to find a rewarding one with someone else.

Every dog must be somewhere in our pack. If we won't responsibly create and maintain a niche where the submissive animal can function confidently and with our full acceptance and support, we deny it our dominion. If we deny it our dominion, we deny the bond between human and canine; and without the bond, there can be no mutually rewarding relationship.

"Well, I'm glad we got Lily straightened out," says Terry as the Pedersons toast their contentedly sleeping canine. "She'll never be a fireball like Rocky, but I can accept that."

"Speaking of Rocky, where did he—" A loud pounding on the back door interrupts the conversation.

"Pederson!" bellows their next-door neighbor. "If that damn dog of yours growls at my boy one more time, I'm going to kick him into the next county and haul you into court!"

In the next chapter we'll explore how owners of dominant dogs may find life even more complicated than those with submissive ones.

6

DOMINANT, INDEPENDENT, AND DESPOTIC DOGS

WHEN THE PEDERSONS' neighbor, Jud Rawlins, accuses Rocky of growling at his son, Terry Pederson downplays the event by noting "Ricky Rawlins is an obnoxious little brat, and everyone knows it. He was probably teasing Rocky, and I can't blame the poor dog for not taking it."

In the flurry of excitement surrounding Terry's promotion to midwest regional sales manager for Trendco Industries, the family forgets the incident. Even though Terry's new job will take him away from home a lot more, he can't pass up this once-in-a-lifetime opportunity to advance in his profession.

"I'm sure glad we have Rocky," comments Terry as he prepares for a two-week trip. "I'll rest easier knowing he's here to protect you."

When Terry returns from his first trip Pamela, Rocky, and Lily rush to greet him.

"Boy, did Rocky ever miss you," Pamela declares later. "He paced almost constantly and barked every time someone came within a half block of the yard."

"That's my boy," beams Terry proudly. "I knew I could trust him." He misses the fleeting troubled look that sweeps across Pamela's face.

When Terry returns from his second trip, he receives an equally joyous greeting, but when Rocky flies happily into his arms, Pamela hangs back. Roughly rubbing the devoted spaniel's ears, Terry peers at his wife. "Rocky take good care of you while I was gone?"

"Too good, I'm afraid," replies Pamela carefully. "Sometimes I think he's overly protective."

"Pamela, be serious! You weigh a hundred pounds soaking wet, live in a populated area with crazies like Rawlins and his kid everywhere. A dog *can't* be too protective!" Terry laughs as he engages Rocky in a mock battle on the living-room floor.

When Terry returns after his third trip, an extended tour that kept him away from home for almost a month, he's surprised, then alarmed when no one greets him. As he pays the cab driver, he can see Rocky leaping joyously in the backyard, his shrill barks surely annoying the neighbors. Why isn't he in the house with Pamela?

As he opens the door, one look at his wife's stricken face and bandaged arm tells him that something is terribly wrong. "We've got to get rid of Rocky," she sobs, shaking uncontrollably. "He almost killed the Rawlins's dog and he even turned on me."

Can a dog owner imagine anything more horrible than the idea that a faithful, devoted, happy-go-lucky pet might some day turn and attack? That's why such events usually make headlines: DOG MAIMS SENIOR CITIZEN, DOBERMAN ATTACKS BABY, BOXER MAULS TEENAGER. Nevertheless, few owners think it can happen to them. When it does, an owner's thoughts and emotions somersault: "The old lady must have harassed the dog." "The dog wasn't properly trained." "The dog must have had a brain tumor." And when it happens to Terry Pederson, Jud Rawlins seems to take great satisfaction in pointing an accusing finger at Rocky: "I told you springers can't be trusted."

If you owned a springer spaniel and listened to Jud Rawlins, you would undoubtedly sense seeds of doubt growing in your heart. Whether they take root and grow eventually to choke the relationship or wither and die depends on your dog's position within your human pack and your acceptance or rejection of that position. As we saw in the previous chapter, a dog's position within the pack and the human response to that position lie at the heart of all problems arising from dominant and submissive displays.

DOMINANT AND INDEPENDENT BODY-LANGUAGE DISPLAYS

We've already mentioned some of the more common dominant canine body-language signals:

- Ears erect.
- Gaze direct.
- Tail up.
- Hackles raised.
- Growling leading to sharp barks.
- Weight shifted to rear legs.
- Leg lifting/territorial marking.

We also noted how a dog will circle and sniff a more submissive animal as well as place its front paws on the other's shoulders to signal dominance.

Again, we must pause to remind ourselves that the dominant individual isn't superior but rather different, possessing qualities that make it the best leader and conversely, and not surprisingly, the worst follower. I think of this every time my husband and I get into one of those typical discussions about who's the best driver. There's no doubt that when we're both in the same car, he's the better driver; but I'm equally sure—and he agrees—that I'm a far better passenger than he. Recognizing our own pet's orientations shouldn't evoke pride or disgust; it should enable us to understand how or where our pets would naturally tend to fit us within their pack structure. Owners of dominant dogs need not be enslaved by their pets any more than owners of submissive animals need lord over theirs. Our awareness allows us simply to understand relative position. Once we know where the dog fits into the human/canine pack structure, then we can decide whether we want to accept that

position or try to change it. For example, had Pamela realized
what Rocky signaled when he insisted on jumping up on her and
placing his front paws on her shoulders, she might either have
discouraged the behavior when he was very young or not become angry and frightened when he growled at her attempts to
push him down when he was eight months old. Unaware of the
meaning of his body-language cues, she initially accepted and
even reinforced Rocky's dominant position. When she unwittingly responded dominantly by attempting to push him down,
Rocky responded to her challenge the only way he knew how—
with a show of more dominance. Had she understood pack behavior, his response would not have surprised her at all and,
more importantly, she could have prepared herself for it.

Independence is a state we confer upon our dogs by virtue
of our emotional interpretation of their behavior in given situations. Many of us use a Goldilocks and the Three Bears system
of evaluation. Whether a dog responds too slow, too fast, or just
right to our way of thinking, there will be an owner who will
view that same response as proof of that dog's independent nature. For example, if Roy Ringhausen commands his beagle,
Chief, to come and the dog casually wanders toward him, goes
off in the opposite direction, or doesn't move at all, Roy might
attribute all these actions to Chief's "independent streak." Similarly, if Chief rushes into a room ahead of Roy or barks at the
ice-cream truck long before his owner even hears it, these actions might also be seen as evidence of the dog's independence.
Finally, if Chief responds perfectly, confidently, and immediately, but not too fast, Roy gushes with pride: "He's one in a
million, isn't he? So together and independent—miles ahead of
the pack!"

While we can accept the validity of all three definitions,
let's combine the "too fast" and "too slow" ones to create a
working definition that can help us compare and contrast independence with dependence. In the previous chapter we defined
dependence as our dogs' unwillingness to do things we believe
they're capable of doing and want them to do. Conversely then,
let's define independence as our dogs' willingness to do things
we believe they're incapable of doing or don't want them to do.
Roy doesn't think Chief can take care of himself; therefore he

wants the pup to be obedient. When the beagle refuses to come when called or trots off in another direction, his owner sees his willingness to behave counter to his owner's wishes as an expression of independence. Similarly when Chief responds too quickly—perhaps even before Roy has a chance to verbalize the command—Roy assumes the dog is deliberately defying its owner's authority.

How does "just right" timing fit into our definition? In terms of our dog's position in our pack, it signifies the opposite of independence. While the dog that doesn't respond at all or responds before we do or without our command is acting independent of our input, the dog that responds to our wishes just as we desire when we desire is depending on our input in order to make its response. Naturally, although a dog can obviously function independent of our beliefs, its willingness to submit to our desires makes it the captivating creature we so dearly love.

Despotic canine body-language expressions are those signals that owners believe a dog exhibits in an attempt to get its own way, and many owners see them as bossy and manipulative. For example, when Chief stands in front of the refrigerator emitting shrill staccato barks, ignoring his owner's commands to be quiet and ceasing only when Mary Ringhausen gives him a slice of cheese, Mary calls Chief's behavior "bossy and mean." Compare this incident with the more passive form of manipulation Richard attributed to Muffin when she refused to eat in his absence. Despotic dogs not only dominate their owners by forcing them to act in a specific way, they can elicit a desired response by preying on their owners' feelings of guilt or pity.

Given these examples, we once again see how the human's position relative to the dog's determines both canine behaviors and owner's perceptions of them.

DOMINANT DOG, DOMINANT OWNER:
TOO MANY CHIEFS

Although Mary Ringhausen shakes her head over Chief's demanding and bossy behavior, their relationship isn't nearly as unstable as the one formed between the dog and her husband. Roy's an outdoorsman at heart, an avid hunter who jokingly tells friends he only works as a stockbroker so that he can afford the best hunting and camping gear. Among his most treasured "gear" is Chief Tamarach of Waubash, purchased from one of the best beagle kennels in the country.

Roy is an independently oriented owner. Much of the excitement he experiences when hunting centers around his attempts to outsmart his prey: "I like to beat them at their own game, no high-powered rifles, no fancy scopes—just me, my dog, and my knowledge of our quarry and the terrain."

In theory Roy should be able to maintain such a relationship with his prey, but in the intimate relationship between human and dog it brings him into acute conflict with his pet. In the wild setting, beliefs like Roy's may add challenge and excitement to the hunt. However, when he demands something from his pet that the latter won't relinquish, the relationship can become a tug-of-war and the household a battlefield.

Watch what happens between Roy and Chief. Chosen because he rushed to the front of the pen and nearly bowled over his littermates in his excitement to jump up on Roy, Chief manifested his dominant inclinations from the very beginning. To further complicate matters, the Ringhausens initially tolerate the behavior for a not-uncommon reason: Because Chief has such impressive papers and cost so much, the owners tend to attribute any shortcomings to their inability to handle such a *rarus canis*. Perhaps such a high-class dog simply doesn't respond to commands the same way Roy's old nonpedigreed hound did.

In the beginning Roy responds patiently to Chief's stubborn refusal to obey unless the mood strikes him; however, with each passing day Roy finds it harder and harder to excuse his dog's disobedience because of his "high-strung breeding." The final blow comes the day Chief ignores Roy's commands to heel and rushes across the highway, barely escaping the wheels of a delivery truck, whose driver squeals to a halt and angrily berates Roy and his "stupid dog."

Enraged, Roy grabs the leash and strikes Chief repeatedly. "Alright, Idiot, that does it. I'm gonna pound some sense into your thick skull." Although Chief gets in a few good bites, the pup's no match for his enraged master. Roy smacks the dog again and again, then drags him home, unmindful of the exhausted pup's gasps for air. Heaving the pup into his pen, Roy slams the door: "Now we'll see who's boss!"

While this scenario may seem somewhat melodramatic, similar clashes occur between dominant dogs and dominant owners in apartments, condominiums, split-level and custom-built homes every day. When owners think their dogs are trying to usurp those powers and the authority they define as uniquely their own, they often resort to fighting fire with fire. Such a solution may give us some momentary satisfaction and even produce apparent results, but it leaves us few alternatives if it fails.

Whenever we let our anger or frustration turn our hands toward violence, it pays to remember the saying, "It's a lot easier to keep a friend than turn an enemy into one." For example, suppose Roy had taken time to cool off, then approached Chief's pen the next day with the idea of a simple training session. Now suppose he commands Chief to heel, but the beagle ignores him. What are Roy's choices? Keeping our four alternatives in mind, suppose, further, that Roy can't accept the behavior or change his feelings about it, but doesn't want to terminate the relationship. That leaves him with one option: changing Chief's behavior. And the only way he can accomplish that goal is to become more dominant, stronger, and more forceful than his dog. Consequently, he beats his dog again, this time harder and longer. Unless Roy can break out of his wild-dog orientation he'll continue this cycle until one of three things occurs:

- He beats his dog into submissive obedience.
- He kills his dog.
- Chief runs away or is taken away by others.

DOMINANT DOG, SUBMISSIVE OWNER: PLAYING SECOND BANANA TO CHIQUITA

Most people who see her laugh at the notion that pint-sized Chiquita is a dominant dog. However, it's not the least bit funny to her owner, Loretta Letesla. Loretta, a widow who lives alone in a retirement community in Saint Petersburg, Florida, received Chiquita as a gift from her daughter and son-in-law: "Something to keep you company, Mom." Unfortunately Loretta never has been and never will be a dog person. In fact, she accepted Chiquita only after a rash of burglaries in her neighborhood made her feel vulnerable: The dog may be a pip-squeak, but her yapping just might be enough to scare off an intruder. Because Loretta isn't emotionally attached to the dog, she makes only the most rudimentary efforts to train her. While she's not cruel to Chiquita, she more or less ignores her during their first six months together, offering praise only when the little dog barks and growls at a strange noise.

Loretta's first inkling of impending disaster comes when she takes the six-month-old Chihuahua to be spayed. Given the dog's natural dominance and untrained status, it's no surprise that Chiquita resists all handling at the vet's. Although the veterinarian suggests that Loretta enroll Chiquita in obedience classes or otherwise train her so that future veterinary care won't be so traumatic for dog and humans alike, Loretta's commitment to the relationship is too minimal to permit her to take this sound advice. Even when such a simple procedure as nail clipping requires tranquilization, muzzling, and manual restraint, even though Chiquita urinates, defecates, and empties her anal glands during the process, Loretta merely chalks up the Chihuahua's behavior to breed temperament.

Even people who enjoy simple life-styles can put themselves through complex gymnastics to eliminate confrontations with a dominant animal. Within a few months Loretta is warning her guests, "Don't sit in the green chair, that's Chiquita's." "Careful, don't walk too close to her or she'll snap." "No, don't touch that toy. Chiquita doesn't like her things moved." Ironically, Loretta's friends interpret her warnings as signs of deep devotion to the dog. Why else would she be so solicitous? In fact, these are ways she's devised to keep out of the dog's way.

We emphasized early in our discussion of canine body language and emotions that only beneficial displays are perpetuated: Those that contribute nothing or cause harm tend to be eliminated. We humans respond similarly: If something isn't working, eventually some crisis forces us to deal with the situation.

For Loretta, it starts out innocently enough. One rainy evening, she's too tired to take Chiquita for her usual long walk. After Loretta's snugly in bed, the Chihuahua nips her toes through the blankets—not a hard bite, just an irritating pinprick. Loretta kicks at the dog, who growls and jumps off the bed, only to return a few minutes later. The sequence repeats itself three more times before Loretta remembers that she failed to walk Chiquita before bedtime and the dog may have to relieve herself. Feeling somewhat guilty, but still in no mood for a late-night stroll in the rain, Loretta crawls out of bed and lets Chiquita out into the backyard. Wonder of wonders, the dog immediately relieves herself and trots quickly back into the house. Seeing a chance to end those daily walks, Loretta praises the Chihuahua profusely and gives her a bit of leftover stew. When Chiquita repeats the display the next night and the next, Loretta repeats the praise and treats until she firmly establishes the nipping behavior.

Soon Chiquita starts biting Loretta on the ankles when she wants something to eat or to go out during the day. Now there are no protective blankets to cushion those sharp little teeth, and Loretta begins sporting Band-Aids on her ankles; if Loretta ignores Chiquita, the dog nips even harder. Soon Loretta becomes a virtual prisoner in her own home. She feels embarrassed and angry that she accepted a pet she never wanted in the first place; she wishes she had paid more attention to the veterinarian's warnings about training; and she's furious with herself for having actually encouraged this bizarre behavior. In light of all these feelings, she can't turn to her daughter, son-in-law, or the veterinarian for help. Furthermore, she becomes so nervous and intimidated around the dog that she stops inviting guests to her home for fear Chiquita might bite one of them. Although she's an attractive woman and loves to wear pretty dresses and go swimming, she begins wearing slacks and heavy shoes and socks even in the hottest weather so that her friends won't see her scabs and scars.

ALL IN THE FAMILY

Mary Ringhausen may have considered Chief too bossy, but she liked the pup and feels a small pang of regret when their relationship terminates after Roy banishes the dog to his pen at the start of the "get tough" training. However, this isn't the case in the Pedersons' household, where Rocky and Lily are such family members that banishment would cause the owners unbearable pain. Interestingly, although Terry and Pamela believe they've been incorporating the dogs into their human pack, certain of their actions suggest that the opposite is more likely true.

For one thing, they both unknowingly relate to Rocky as if they themselves were dogs, yet in entirely different ways. Terry crouches down on all fours and "growls" at Rocky, grabs him by the shoulders and flips him over onto his side—the classic alpha roll exhibited by dominant canines. While Terry inadvertently establishes his dominance in this way, Pamela sends Rocky signals of submission. For example, the few times she tries to imitate Terry's rough play with the spaniel, Rocky becomes so excited, he bowls her over. Although he doesn't hurt her and seems perfectly friendly, it seems to her he's always trying to get the upper hand, standing over her when she lies on the floor, putting his front paws on her back or shoulders whenever possible. And while she enjoyed and even encouraged this behavior when he was a fluffy pup, it bothers her as he grows older and becomes more demanding. By the time Rocky reaches seven months of age, Pamela never sits or lies on the floor near the dog and, in spite of Terry's protests, she refuses to allow the dog to sleep in their bedroom. Although she tells her husband she objects to the dog's shedding, in reality she finds Rocky's presence increasingly intimidating. In fact, she would love to invite gentle Lily to sleep on the bed and feels resentful that Rocky's bossiness makes this impossible.

One day Pamela attempts to move Rocky's food dish, and when the spaniel growls at her, she jumps back immediately. Although she scolds Rocky, she's sufficiently startled and frightened by the behavior that she can't muster much authority in her voice; her admonition sounds so much like a submissive whine that Rocky ignores it.

Rounding out the Pedersons' irregular pack behaviors are those the family uses to fulfill their definition of Rocky as a protector. Here again, Terry and Pamela differ in their opinions of acceptable protective behavior and reinforce different displays. Because Terry sees protecting his household as a paramount concern involving force if necessary, he reinforces strong physical displays in his dog. When Rocky lunges and snarls at the Rawlinses' Doberman as she walks past the Pedersons' house with Ricky, Terry says, "Good boy," not because he fears the Doberman or the boy but because he believes the Doberman might threaten Pamela when he's not there to protect her. Recognizing there's no need for Rocky to attack, he then aborts the display by commanding his dog to sit and stay.

While Terry's actions may seem consistent, further analysis reveals that he and Pamela have conspired to create a relationship with Rocky that gravitates relentlessly toward the eventual disastrous homecoming. First, they have a dominant dog who recognizes only one family member as more dominant. (How often have you heard one family member, usually the more timid person in the household, say "I don't know what's wrong with that dog. He'll do anything Frank commands, but acts as if I'm not even here"?) Second, one family member responds submissively, backing off and "whining" whenever the dog challenges her. And third, the dominant member creates and reinforces a specific sequence of events: Terry praises Rocky for a defensive display, then asserts his own dominance by ordering the dog to abort the display.

Add to these characteristics of the relationship the fact that the submissive person can't effectively carry out the sequence established by the more dominant one, and the stage is set for conflict. Sometimes two owners with differing pack roles don't even realize such a sequence exists, often because the dominant one simply assumes that the other knows about it or makes the even graver error of assuming that whatever the dog does for him, it will do for someone else. Terry does believe that anything Rocky does for him he'll do for Pamela, but his belief relies upon two fallacies. In fact, Rocky *wouldn't* submit to Pamela's authority the way he does to Terry's. And even if he would, Pamela would be afraid to ask him to do so.

Furthermore, Terry erroneously thinks Rocky will respond negatively only to those people or events Terry himself considers threatening. He maintains this belief because Rocky's behavior in Terry's presence seems to support it. However, he forgets that Rocky takes his cues from Terry: If Terry responds positively to another so does Rocky, or at least he's willing to abort any unshared negative response at Terry's command. But what happens when Terry's not there to deliver the cues?

For her part, Pamela knows from bitter experience that Rocky can be quite a different animal in her husband's absence; but the few times she raises the subject, Terry admonishes her to be more firm with the dog. He then proceeds to demonstrate, and Rocky performs flawlessly, making Pamela feel more inadequate than ever. Before the Pedersons' pups are six months old, Pamela calls Rocky "Terry's dog" and Lily her own.

This pack structure serves the Pedersons well enough, provided Terry stays home and Pamela squelches her negative emotions: "Grow up, Pam!" she scolds herself. "It's stupid to be jealous and afraid of a basically good pup." However, when Terry accepts the regional sales position, the pack structure begins to alter subtly and ominously.

The first time Terry goes away, Rocky's willingness to take over the household causes little distress. It rains most of the week, and few visitors or interruptions create any clashes in authority between Pamela and Rocky. The few times Rocky rushes to the door and barks, Pamela considers it appropriate behavior and praises him. Because this is the first time she's been alone in the house for an extended period, she really does find Rocky's vigilance a great comfort, although she also begins to feel mildly uneasy about his failure to obey her commands to stop barking.

The second time Terry travels on business, the situation deteriorates markedly. Not only does Terry take a longer trip, the pleasant Indian-summer weather seems to bring everyone and everything outdoors. Rocky divides his time between barking at anyone who comes near the house and lunging at bicycles, pedestrians, and other dogs. Pamela's arms ache so badly from trying to restrain him during a brief walk, she decides to limit Rocky's outdoor activity to the fenced-in yard. When he remains indoors, he paces and whines almost constantly, periodically uttering sharp alarm barks, which make Pamela jump and rush to the window or door. Sometimes his patrols continue throughout the night, making sleep impossible. One night Pamela is awakened from a troubled sleep by the ringing phone: "If you can't keep that damn dog quiet, I'm gonna call the cops!" roars an unidentified caller. "That does it," she swears as she stumbles out of bed and makes her way to the kitchen. Even though she knows she'll never get it away from him again, even though she knows it will give him diarrhea, Pamela tosses Rocky a big leftover hambone to distract him for a few hours.

During Terry's extended trip, the tension finally erupts into a full-blown crisis. By now Pamela's relationship with Rocky is infected with a host of negative emotions:

- She's afraid of him.
- She resents his bossiness.
- She's jealous of his relationship with Terry.
- She feels guilty about not liking him.
- She feels stupid because she can't make him obey.
- She's frustrated and angry at Terry and Rocky for putting her in this horrible position.

The day of reckoning begins innocently enough with Pamela making preparations for one of her favorite holidays, Halloween. Having been an area elementary teacher for several years, she enjoys seeing former students and sharing the excitement of the younger children. Sporting a witch's costume, she answers the first knock at the door, only to find ten-year-old Ricky Rawlins and his devoted Doberman, Heidi, waiting for a treat.

Though we already know the end of the story, let's dissect how and why it happened. First, Heidi represents an "evil" presence against which Terry encouraged Rocky's attack/abort sequence. Second, Rocky views adolescent Ricky as a potential competitor with whom he's already suffered some frustrating encounters. Third, even though Heidi's not a particularly dominant animal, she is totally devoted to young Ricky. Finally, we know that Pamela can't control Rocky.

When Pamela opens the door with Rocky at her side but slightly ahead of her, the dog sees Ricky and growls. The boy screams and jumps back. Heidi responds to her master's fear by lunging at the springer. When Pamela tries to grab Rocky's collar, her dog glares at her, snarls, and almost casually rips her arm. Spurred on by Ricky's hysterical screams, the Doberman fights valiantly; but Rocky has the advantage from the beginning because:

- He's on his own territory.
- He's much more dominant than Heidi.
- He's been conditioned to make this response.

While the drama seems to unfold in agonizingly slow motion to Pamela, the actual event lasts mere seconds. Heidi is rushed to the vet's, Pamela to the emergency room. Although Jud Rawlins would have liked to kill Rocky then and there, he simply drags the unresisting dog into the Pedersons' backyard and locks him in. There Rocky remains until Terry returns three days later.

GETTING BACK ON THE RIGHT TRACK

Whether we consider Chief's stubbornness, Chiquita's tyranny, or Rocky's viciousness, all three dogs and their owners share the same problem: Neither Roy, Loretta, Pamela, nor Terry have established dominion over their pets. With their forceful approaches, Roy and Terry may achieve dominance, but never true dominion. And while they may be able to guarantee their dogs' responses to themselves, they can't guarantee that their pets will respond similarly to others. A one-man dog philosophy may nourish an insecure ego, but most dogs in such situations actually suffer from a lack of training. If the one-man response includes protective or other potentially dangerous displays, the dog is even more behaviorally unstable than a completely untrained one. Because such owners base relationships on the presence of their own greater force, their dogs receive two clear messages:

- Force is how you get others to do things.
- If that force isn't present, you can do whatever you want.

Consequently an approach based on outdominating the dominant dog often reinforces the dog's dominance toward all but one person or people like him or her; and it rules out the possibility that others will be able to enjoy a rewarding relationship with the pet.

Owners who cultivate such relationships can expect to grapple with some painful introspection when problems arise. For example, if Roy's methods turn Chief into a remarkable hunting dog, he'll feel understandably proud and confident of those methods. However if Chief refuses to respond to his best friend and hunting companion, Roy's reactions are more troublesome. On the one hand he might rationalize that he's the better trainer and handler and that Chief's so devoted to him he won't work with anyone else. On the other, he might remember how great he felt when his friend's dog responded flawlessly to his own less-than-perfect handling in the past.

Terry must also come to terms with the fact that he reinforces much of Rocky's protective behavior because he likes the feeling of power and control he feels when he can abort a hostile display with a simple command. He never really wanted Rocky to attack anyone or anything. In fact, when Pamela tells him what occurred, he shrinks back from his faithful dog with repulsion and disgust.

Meanwhile, by submitting to their dominant dogs' displays, Loretta and Pamela also shirk the responsibilities that accompany dominion. However, whereas we can cite concrete acts Roy and Terry used to establish dominance, Loretta and Pamela signal their submission with much more subtle body-language cues. Most often they passively show their submissive rank by staying out of the dogs' way. Loretta begins snacking rather than eating regular meals to avoid the mealtime biting; she sleeps in the tiny guest room because she can more easily keep Chiquita out of there at night. Pamela keeps Rocky outdoors as much as possible when she's home alone. In such ways submissive owners avoid confrontations with their pets and perhaps even develop a false sense of security because they "won that round." However, only direct confrontation, not avoidance, can ever really change the relationship.

Other times submissive owners more actively signal their lower position. Every time Loretta feeds, walks, or otherwise responds to Chiquita's nipping, she accepts Chiquita's dominance *and* rewards her for it. Every time Pamela backs off when Rocky growls, allows him to jump up or do anything she doesn't like because she's afraid to stop him, she also rewards his dominance.

From these examples we can see that lack of dominion hurts submissive owners just as much as it hurts dominant ones. Furthermore, we can see that only self-confidence can get the relationship back on the right track. While Terry and Roy may feel much more confident about the nature of their relationships with their pets, the fact remains that these relationships spring from their *lack* of confidence. Only by developing confidence in his ability to teach Chief and in Chief's ability and willingness to learn can Roy abandon the use of force; and only by truly believing that Rocky loves him and Pamela enough to protect them *when it's necessary* can Terry avoid reinforcing hostile displays toward nonthreatening elements. Owners who cry wolf to test their pet's obedience and loyalty risk creating a beast on whose reactions they can never confidently rely.

Submissive owners lack confidence of a different sort. Whereas dominant and independent owners believe dogs won't learn or behave properly unless forced, submissive owners believe this *and* that they lack the force necessary to produce the proper relationship. Loretta believes that if she could find the courage to knock Chiquita up the side of the head, she could solve all her problems; Pamela believes Terry's relationship with Rocky outshines her own. However, by simply adopting a more dominant approach, both women would accomplish only the exchange of one kind of problem relationship for another.

RESOLVING THE DOMINANCE DILEMMA

When problems arise with a dominant dog, one of the biggest confidence builders comes from evaluating the four options. Nowhere does commitment to the dog become more critical than in a confrontation of violent behavior, whether it's the owner's, such as Roy's heavy-handed training, or the dog's, such as Chiquita's nipping and Rocky's biting. Without the confidence that arises from the knowledge that we want to keep the dog and build a good relationship, we're defeated before we even begin.

For example, Loretta didn't want Chiquita from the beginning. Although she feels embarrassed by her contribution to Chiquita's negative behavior, her reasons for altering it have nothing to do with wanting a sound relationship with her pet. She doesn't want any dog, but if circumstances force one on her, then she wants one that doesn't bother her and, most certainly, one that doesn't complicate her life in any way.

Because we know that the solution to dominant-dog problems begins with commitment to the relationship, confidence, and dominion, what are Loretta's chances of establishing a stable relationship with Chiquita? Not very good. In fact, when she faces the four options squarely, Loretta realizes the only reason she would bother changing Chiquita's behavior would be so she could ignore her again. Furthermore, if she had her choice, she'd get rid of the dog.

Fortunately, Loretta gets in touch with a qualified trainer-counselor who helps her see that she can make a choice without offending anyone and that she should shoulder the responsibility for choosing either to deal with the situation or terminate the relationship.

"But what will my daughter and son-in-law think? Surely they'll think I'm ungrateful."

"Ms. Letesla, do you honestly believe your family wants you to live with a dog you dislike who imprisons you in your own home?"

Although Loretta shakes her head no, her embarrassment and guilt regarding what she views as her own culpability make it impossible for her to discuss the matter with her family. To get around this obstacle, the trainer suggests that he contact Loretta's daughter for her, simply offering his professional opinion based on his knowledge of both owner and pet. After Loretta agrees, the trainer's suspicions are confirmed: Loretta's family had no idea the dog was causing such hardship and readily agree a new home should be found for Chiquita following some professional training. The trainer agrees to keep the dog until this is accomplished.

As we might expect, Roy and the Pedersons immediately and emphatically rule out termination. Even though Pamela has been badly intimidated and hurt by Rocky, she wants to keep him because she knows how much he means to Terry. Both men realize they prize their dogs enough to want to do whatever they can to establish better relationships between their dogs and themselves, their families, and others.

Let's look at Roy and Chief first. Roy isn't afraid of Chief and believes the beagle is intelligent and capable of learning. However, he also believes Chief doesn't want to learn and won't learn without force. Because of Chief's dominant nature, the more Roy forces him, the more stubborn the beagle becomes. Roy needs to loosen up a bit, working with his dog instead of against him. No, he needn't "give in" to his dog, but he should set a new goal—the creation of a well-trained hunting dog who enjoys his work, rather than a canine robot who responds to any command, anytime, anywhere. The former, rigid approach to training might work and even be necessary to give confidence to highly submissive or fearful animals. However, intelligent, more dominant, goal-oriented dogs like Chief don't need, nor do they respond well to, that process. Chief was born and bred to hunt rabbits. When Roy attempts to train him using a "Jump when I say jump!" approach, we can almost see the pup trying to figure out what that has to do with rabbits. When the process becomes more important to Roy than the goal, when he worries more about who "wins" the battle of wills rather than whether or not he and Chief enjoy hunting together, the relationship can only deteriorate.

By training Chief in a way that allows the dog to succeed, Roy's appreciation for the beagle's high intelligence and willingness to please skyrockets. Instead of placing a lead on the pup, staring him straight in the eye, yanking the lead, and demanding that Chief come, Roy gives the command as he himself begins running in the desired direction. Although Chief initially holds back and Roy feels a slight drag on the lead, he keeps moving. Soon he's delighted to notice the pup moving easily at his side. In this way Roy gets his dog to do what he wants him to do without getting bogged down in the tug-of-war his previously dominant approach always precipitated.

While some may argue against this as an orthodox training procedure, once we use such techniques to establish a solid relationship with a dominant dog, we can become process-oriented again if we wish. If we ask a dog to do something that violates its breeding or experience, we need to provide a good reason for it to do so. Nothing in the wild dog's evolution makes it advantageous to come, sit, stay, roll over, or shake hands in response to human commands, to say nothing of sniffing out or retrieving prey that it isn't allowed to eat. We must supply that reason, and the only workable one is that the dog wants to please us. Attempting to force a response from a dominant dog takes a lot of time, effort, and strength and often produces a spiritless machine-dog or a barely controlled disaster.

Having saved Chief from life as a machine-dog, let's see if we can help resolve the Pedersons' calamity. Terry isn't afraid of Rocky and has confidence in Rocky's intelligence as well as the spaniel's willingness to learn. However, the biting incident shows him how his relationship with Rocky increased his dog's protectiveness and contributed to Rocky's belief that he was in charge of the household in Terry's absence.

The first thing Terry does is stop relating to Rocky like another dog would, replacing the roughhousing with simple commands and fetch games. While Terry doesn't treat Rocky meanly, he does begin ignoring the many bossy behaviors he used to reinforce, often unknowingly. For example, when Terry or Pamela were reading the paper or involved in other activities and Rocky wanted attention, the dog would stick his face in their laps or even jump on them. They usually stopped what they were doing, laughed, and petted or scolded him. Either way, they rewarded the behavior by giving Rocky what he wanted: attention. Now whenever Terry sees Rocky heading toward him, he either gets up and walks away or commands the dog to sit; then he approaches the dog and praises him for the proper response. In such a way, Rocky learns to earn rather than demand attention. Terry also discourages Rocky's jumping behavior, using a similar combination of turning away or distracting Rocky with commands for more appropriate, less dominant displays.

Because of the seriousness of the Halloween incident, Terry also enrolls Rocky in a group obedience class so that he can work around other dogs. Realizing that a dog that has attacked another—regardless of why—may not be welcome in certain classes and could be dangerous, Terry discusses the spaniel's problems in depth with the trainer beforehand. Indeed the first trainer he consults tells him that her classes are for novice owners with stable dogs, and she can't afford to include a biting dog in a large mixed group. But she recommends a colleague who does work with problem behaviors, and Rocky and Terry soon become part of an elite group of eight with dominance problems.

The trainer begins with individual weekly sessions, which expand the basic training regimen Terry has already initiated at home. Once owners are confident they can handle their dogs at home, they work together as a group. Soon owners can actually trade dogs and work one another's with equal success. Animals that previously felt they had to call all the shots recognize the dominion of their owners and others and become much more stable and confident.

DOMINANCE AND DOMINION: A BONDED VIEW

Many owners and even some trainers and veterinarians equate dominance with demanding, bossy dogs. If so, they see the dominant dog as harder to handle, more likely to run roughshod over the owner, and more likely to become hostile; and they assume that dogs with naturally more dominant personalities must be forced into a submissive position within the human pack. However, we've seen how many of the problems associated with rank occur not only as a result of a dog's dominant or submissive behavior, but as a result of those behaviors playing themselves out against a backdrop of circumstances and owner orientations. Matched with an owner who refuses to accept dominion over his or her pet, any dog—regardless of age, rank, sex, or serial number—can easily run amok and make life miserable for everyone around it.

Bonded owners don't look at dominance as a fixed characteristic that permits them to stereotype their pet or excuse certain behaviors. Rather, they use their awareness of their pet's and their own relative rank to determine who's in charge whenever problems arise. When Terry learns the difference between dominance and dominion, he realizes he expected far too much from Rocky in his absence. How could he expect the spaniel to differentiate instinctively between threatening and nonthreatening stimuli? Recall how Cathy McCaffrey made this same error with submissive Czar in the previous chapter. In both situations, the owners literally armed their dogs without giving them any knowledge of the rules governing legitimate targets and the proper use of their weapons.

Dominant dogs usually respond faster and more actively than their more submissive colleagues. That very quality attracted Roy to Chief the first time he saw the dog, and that extra bounce made Rocky a more exciting pet to Terry. However, that characteristic also demands that we have dominion over our dominant dogs as much as, if not more than, over their submissive cousins. Just as we exert our dominion to impart confidence and decrease dependence in the submissive animal, we can also use it to channel the more independent nature of the dominant canine properly. Without such guidance the dominant animal has the potential to get into problems faster than others, an all-too-frequent occurrence when we ignore the implications of pack behavior in our relationship with our pet.

Above all, bonded owners recognize that dominance no

more equals bravery than submission equals cowardice. Recognizing dominance as nothing more than relative rank, we can appreciate how a dog may respond dominantly in one situation and submissively in another. When Chief's littermate accepts his dominant displays submissively, no fight erupts; nor does one occur when Chief responds obediently to Roy's commands. A bonded Roy can quickly run toward Chief's excited "Come see what I found!" howls without feeling the least bit abused or dominated by his dog.

By now we should understand that dominance in and of itself isn't a problem at all: The problem stems from how we respond to it. Given an understanding of dominance, sufficient confidence in ourselves and our dominant pets, and willingness to accept dominion over them, we can enjoy these animals who add their own very special zest for life to any bonded relationship.

Haven't we forgotten something? Are we simply going to abandon Pamela with her bandaged arm and ambivalent love-hate relationship with Rocky? Indeed not: The beliefs and problems associated with fear and aggression are so complex, we're going to devote an entire chapter to them. Nothing can so devastate a relationship between human and canine; and invariably this fear strikes when one views the other's behavior as "aggressive." However, even though most of us can fairly quickly recognize these emotional states in ourselves, we have to concentrate to discern them clearly in our dogs. Practically everyone who heard the story of Rocky and Pamela said, "I knew that would happen," yet the biting incident took Pamela completely by surprise, a realization that increases her fear even more. Is it possible she's blind to such negative traits so obvious to others? Or is Rocky so devious that he successfully hid his true nature from her until he saw a chance to strike?

Of such questions are dog owners' nightmares made. Let's conclude our discussion of submission, dominance, and dominion by dragging the worst of the behavioral and emotional skeletons out of the closet: fear and aggression.

7
FEAR AND AGGRESSION: THE BEST AND THE WORST

TERRY PEDERSON BECOMES so engrossed in Rocky's training that several weeks pass before he realizes Pamela barely even looks at the dog. Although he initially understands her reluctance to work with Rocky until her arm heals, he can't comprehend it once she's fully healed several weeks later. Why won't she at least play ball with Rocky?

"I can't Terry, I just can't. I don't trust him."

"But he's not vicious. We just hadn't trained him properly. Trust me, all you have to do is work with him. You'll see."

"You don't understand. He attacked me. How can I trust a dog that attacked me?"

"Pam, you're forgetting what the trainer said. Rocky didn't attack you specifically. You just happened to get in his way."

Rather than become embroiled in an argument, Terry takes Rocky for a long walk. While they're gone, a deliveryman comes to the door, and to Pamela's chagrin, the ever-calm Lily leaps off the couch and begins barking furiously.

"Lily! What's the matter with you?" shrieks Pamela. "Don't tell me you're going to turn into a vicious beast like your brother!"

Across town Howie Chan's family feels equally exasperated

with their dog's behavior. Their proud and elegant eight-month-old Akita, Kimo, turns into a quivering mass of canine Jell-O every time he hears thunder; and during the last six months the problem has grown progressively worse.

"By the time he's a year old, we won't be able to haul him out from under our bed on a cloudy day!" groans Howie as Kimo slinks toward the bedroom during a particularly fierce storm. Howie picks up one of his son's toy trucks and pitches it to the floor just in front of the fleeing canine. Kimo leaps in surprise, defecating, urinating, and emptying his anal glands in the process. "Alright, that does it—now you're really going to get it!" roars Howie as he grabs the terrified dog by the scruff of the neck.

Connie Chan shakes her head in disbelief. "Howie, stop it! You're scaring Kimo to death." As usual Howie ignores her plea and begins shaking and spanking the dog. Regardless of what Kimo does, his behavior never satisfies Howie. Even if the Akita does the right thing, he never does it fast enough for Howie; if he does manage to do it very quickly, Howie's likely to fault him for being too fast. To Connie it seems as if her husband were actually competing with his own dog in a game where Howie periodically changes the rules to make sure Kimo always loses.

Of all the emotional and behavioral states that affect the canine-human relationship, fear and aggression are the two most misunderstood and maligned. While we generally consider these states undesirable and seek to avoid them at all costs, in reality they both serve protective functions critical for the survival of the individual and the species. Given this fact, why does the idea of a fearful or aggressive dog strike us as abnormal, cowardly, or even dangerous? Why does the thought that we might be afraid of our dog alarm us so much? Why do we shudder at the image of ourselves aggressively beating a frightened dog senseless? The answers lie in the differences between fear and submission and between aggression and dominance. But before we can fully appreciate them, we must differentiate between fear and aggression themselves. Let's begin with an overview of some of the basic facts behaviorists propose.

FEARFUL AND AGGRESSIVE SIGNALS

To properly understand fear and aggression, behaviorists needed to know what, if anything, differentiates fear from anger, because most people use the words *anger* and *aggression* synonymously. From the behaviorists' point of view, if fear and anger produce identical behavioral changes, then fear and aggression must do likewise; if not, we must use those terms more carefully.

As it turns out, the conventional understanding doesn't withstand the test of science. Fear and anger have specific and quite different effects on the gastrointestinal tract and other involuntary functions. When an animal feels anger, its gastric and intestinal motility slows significantly; its mouth becomes dry, its digestion grinds to a halt. Physiologically this phenomenon permits the shunting of energy away from nonessential maintenance functions and toward the voluntary muscles needed to fight or flee. Fear produces the opposite effect: Gastric and intestinal motility increase, saliva flows freely, butterflies flutter in the stomach, and nausea and diarrhea make it difficult for the frightened individual to do anything but cower in place. In other words, fear favors the freeze response.

Scientists also relate fear to early isolation. Puppies raised in total isolation won't respond to stimuli as potentially damaging as electrical shocks and open flames because they lack the fear that repulses and protects their socialized counterparts from such dangers. Therefore, it appears that total isolation inhibits fear reactions to a point detrimental to an animal's well-being.

Limited isolation produces a different effect. Pups confined to a small area with minimal human or animal interaction during the critical socialization period that occurs between five and twelve weeks of age and beyond exhibit behavior unlike that displayed by completely isolated ones. Partially isolated pups develop such a meager worldview based on their early experience that almost anything they encounter outside that limited environment elicits fear; and anything that violates the spatial limits of that early kennel-world appears threatening. I learned a lot about this phenomenon from Tully, a dachshund who spent the first year of her life in a kennel as part of a breeding colony. When placed in the center of a room, Tully immediately looked for some kind of enclosure in which to hide, zipping under tables or chairs with lightning speed. If she saw no shelter, she would keenly observe every movement in the room. Whenever anyone or anything approached the limits of her mental "kennel," she began showing unmistakable signs of fear.

Such observations lead us toward two conclusions. First, the fear response apparently physiologically predisposes an individual to freeze rather than flee or fight when encountering a threatening situation. If we think about typical pack interactions, this makes sense because the more submissive animal—the fearful one—should freeze to avoid attack. Thus its physiological response reinforces the proper behavioral one.

Second, the relationship between isolation and the fear response strongly suggests that fear depends on an individual's *relationships* with others and the environment rather than on genetics or instincts. In fact, most scientists believe that the fear of falling is the only innate fear; all others are believed to be learned.

This brings us to another phenomenon, the development of specific fears as a function of time and experience. Behaviorists refer to the period between eight and ten weeks of age as the fear-imprint stage because experiences the pup perceives as fearful during this period may precipitate fears that persist throughout the dog's entire life. Who would have thought that the Chan twins' sneak attack on ten-week-old Kimo with a cherry bomb would result in his turning tail and running every time he hears thunder?

Few people pay much attention to the many ramifications of the fear-imprint period, during which:

- Most pups are sold or given away.
- Most pups experience their first encounter with a veterinarian.
- Most owners initiate housebreaking and other training.

Thus, during a period when we should be providing the most stable and positive environment for our pet, we subject it to a great deal of change and trauma. What a testimony to the dog's basic stability and the strength of the human/canine bond that the majority survive our ignorance so well!

Do young pups exhibit any aggressive responses comparable to those early fear displays? While behaviorists don't classify it as an innate aggressive response, neonatal pups will respond defensively to pressure on the neck area, attempting to pull away from it (flee) or bite (fight) it. We recognize the value of this as a protective mechanism to ward off predators. However, because bitches routinely transport their pups this way, wouldn't such a purely defensive response be self-defeating, complicating necessary relocation? Not if pups only respond defensively to foreign pressures (that is, all but their mother's). If so, the behavior would be beneficial and would consequently be perpetuated. In fact, experimental research supports this possibility: The very strong aggressive "do something" response is tempered by the learned "don't move." Moreover, the two systems work together to provide the optimum protection for the helpless pup:

- The instinctive pull away or fight response enables even the youngest to mount some defense when alone.
- The learned freeze response allows the bitch to move the pups with maximum efficiency and minimal noise and resistance.

We needn't stretch our imaginations far to realize that the innate defensive response confers a survival advantage and competitive edge. In other words, it demonstrates the animal's aggressiveness, its desire to survive, and its willingness to do whatever it can to achieve that goal. This antagonism toward the alien is balanced by the learned respect for authority inherent in the freeze-response signals. In both situations the physiology supports the response. The hypermotility of fear reinforces the animal's instinctive impulse not to move; the shutdown of nonessential involuntary functions makes more energy available to flee or fight when survival depends on action.

Putting it altogether, we can begin to appreciate the delicate balance between fear and aggression and marvel at how exquisitely it provides an individual with three distinct mechanisms for survival. Which display the dog makes depends on its evaluation of its relationship to the threat. Is it more advantageous to move away from it (flee), stay put (freeze), or move toward it (fight)? Or even more basically, should the dog do something or nothing?

If the animal feels it can't possibly survive either by dominating the threat or submitting to it, it flees. If it feels it can dominate the threat, it fights. If it feels fleeing or fighting are impossible or unnecessary, it freezes.

To some extent animals also employ the freeze display to signal stronger, nonthreatening individuals that outside help would be appreciated. The pup who freezes in response to a threat makes the connection "This threat is something Mom should move me away from. I'll hold still to help her do so." In such a way, the freeze response serves two functions:

- In response to members of the same species (or perceived same species, such as humans incorporated into the dog's pack), it aborts any attack.
- In response to something the dog believes beyond its ability to fight or escape, it pleads for outside help.

In such ways the active flee-or-fight and passive-freeze responses offer the animal a full range of options. While this ingenious balance of active and passive alternatives works beautifully to preserve the integrity of the individual wild dog and the pack, it doesn't fare so well in the human/dog pack. In the wild assemblage everyone understands both the signals and the appropriate responses; in the domestic conglomerate, dog and human don't always adhere to the same rules and frequently don't even play the same game!

In summary, when a dog encounters events involving others, it immediately evaluates the other as friend or foe as well as dominant or submissive in terms of itself. If the other is perceived as a threat, the dog's perception of its relative dominant or submissive position determines whether it will respond aggressively ("do something") or fearfully ("do nothing"). If we can learn to recognize these body-language expressions, we can determine how the dog perceives us and the situation as well as what it might do. By understanding how the dog selects from its available options we can avoid inadvertently precipitating or contributing to unnecessary and even harmful negative responses.

LEARNING THE BASICS: MAKING THE RIGHT CHOICE

Obviously as a pup grows, its learned experiences begin to color even the strongest intuitive responses. While animals experiencing a threatening event intuitively choose whether to flee, freeze, or fight, they make such choices individually according to circumstances. They also develop habitual responses. The fleeing animal who survives will probably flee again when similar circumstances arise, because the chosen body-language display helped ensure its survival and that of the species.

Furthermore, the flee, freeze, and fight choices offer a hierarchy of response ranging from minimal to maximal contact. To see how this works, let's look at a commonly perceived threat: loud noises. We already know that Kimo reacts with terror to such sounds. Suppose the same sounds intimidate but don't terrify Lily, whereas Rocky associates them with the obnoxious Ricky Rawlins. If we were to sneak up behind Kimo, Lily, and Rocky and make duplicate offending noises, Kimo would dive for cover, Lily would freeze, and Rocky would whip around and snap. In the presence of this particular stimulus, Kimo feels the most threatened and reacts to the noise as if it were life-threatening, choosing the route that promises the least contact with his adversary. Although Lily doesn't particularly like loud noises, she doesn't find them life-threatening, so she freezes. She's confident from her own experiences that if she submits—freezes with ears back, eyes averted, tail tucked, and legs rigid—the threatening presence will either leave her alone or someone (Rocky or the Pedersons) will make it go away.

Rocky doesn't like the noise and does feel threatened by it, but he believes he can subdue it; therefore he makes a stand quite unlike his sister's. His direct gaze, high tail, and body balanced to spring send an obvious message: Rocky thinks he can dominate the situation and will prove it if necessary.

The flee, freeze, and fight options also form a hierarchy within each individual as well as the population. When he hears thunder, Kimo first attempts to run. However, if something or someone blocks his escape—Howie stands in his way, for example—he may freeze if he perceives the obstacle as sufficiently threatening. If his fear and desire to run increase intolerably, he may even be willing to fight, snapping at Howie if necessary to get past him to the bed.

The hierarchy of responses also appears when a fleeing animal is cornered. It wants to run but can't. Next it may freeze in hopes the threatening individual or situation will go away and leave it alone. If the adversary also ignores this display, or if the cornered animal feels so threatened that freezing offers no apparent advantage, it fights.

Compare this last-resort fight response to that made by an animal who believes it can dominate a given threat. While the latter display may seem like an "unprovoked attack," both animals provide distinct and predictable body-language cues to someone aware of their meaning as well as of his or her own relationship to the dog. People who are bitten almost invariably ignore or don't recognize the displays the animal uses to signal that it considers the person or the situation a threat it wants to avoid, submit to, or dominate. Not only don't they recognize the dog's message, they don't realize that the dog expects a specific response from them as a signal of their acceptance or rejection of its position.

Even those who do recognize that they appear threatening to a dog often precipitate disaster because they take the dog's evaluation quite personally (and emotionally) and seek to change it instantly. Surely every owner of a more submissive dog has indulged guests who were determined to "make friends with the dog," even though the dog clearly found them threatening for some reason. The dinner gets cold, the business goes unattended while Harry pursues Curry through the house and around the yard: "C'mon, fella, I won't hurt you." Probably not, but Curry's never seen a six-foot-two-inch bearded guy wearing glasses, cowboy boots, and an earring, and it's going to take time for him to accept such an apparition as a normal, let alone friendly, addition to his world. Friendship would occur much faster if Harry would just leave the dog alone.

However, if Harry thinks of himself as a lovable guy and a dog fancier to boot, he probably won't leave the dog alone. When Curry resists his advances, he's inundated by embarrassment and guilt. Because he believes Curry's fearful displays reflect negatively on him, he also becomes angry and resentful. As the dog's persistent wariness begins to irk him, his "C'mon, dog, don't be such a chicken!" gives way to "Pleeeease, don't be so stubborn!" with a hard edge to the words. Obviously this only serves to intimidate the dog and enhance the display even more. Eventually if the owner is lucky, Harry gives up, noting, "There's something wrong with your dog, buddy." If the owner isn't lucky, Harry continues pursuing the poor creature until he corners it. Only his yowl of pain and furious "That vicious mutt of yours bit me!" will terminate the confrontation.

AGGRESSION, FEAR, AND CHANGE

From these examples, we could erroneously conclude that aggression plays a role in the survival of the animal only when it's confronted by a threat. However, aggression functions in other ways, too. As we noted before: A response must not only be good, it must be good for *something*. By the same token, functioning at mere survival level means that the individual or species maintains itself on only the most basic level. To evolve or adapt to changing circumstances, individuals must be willing to change. This willingness can also be defined as a form of aggression.

Think of aggression as a competitive edge, the willingness of one individual to respond faster, longer, or more strongly to achieve a particular goal. Thus within the canine species, for example, aggression depends on the breed, the individual within the breed, and the situation the individual encounters, among other variables. Because of their heightened visual sensitivity to motion, the Pedersons' spaniels respond much more aggressively to birds in flight than the Doberman next door; however, even their response pales beside that of a sight-sensitive greyhound. A beagle's aggressive response to scent makes the average Pekingese look like an olfactory novice, but a bloodhound can outsniff a beagle any day.

Individual differences such as sex and age also affect aggression. As the more dominant littermate, Rocky is more strongly territorial than Lily. Consequently, he responds more aggressively—faster, longer, more forcefully—to anything that encroaches on his territory. Not only does he challenge threatening intruders more actively than Lily, he also greets nonthreatening ones more enthusiastically.

When Rocky and Terry leave the house, however, the purpose—the canine *raison d'être*, protecting the territory—falls on Lily, and this rare responsibility may affect her behavior dramatically. Whereas she normally remains on her rug by the woodstove and yips only once or twice (and rarely ever *before* Rocky), when Rocky's gone, she reacts just as strongly and quickly as he to changes in the environment. Similarly, while shy females may routinely yield to dominant displays from other dogs or humans, they may respond quite aggressively when nursing pups.

Consequently, the potential to respond aggressively exists in both dominant and submissive animals. How or if either type actually expresses aggressive behavior depends on the strength of the underlying purpose and the animal's relationship to the threat and its environment. So while Lily accepts maintaining her territory as part of her canine purpose, she also recognizes that as long as Rocky's more willing to make the necessary displays, there's no reason for her to do likewise. In fact, doing so would actually challenge him and create friction rather than stability in their pack. Therefore, in Rocky's presence Lily's displays take a mild form; but when he leaves she grows much more willing to display her position as territorial protector physically, because that's now a beneficial and necessary stance.

While she may not respond exactly like Rocky—there's no way she's going to rush up and growl at the garbageman, for example—like Rocky she sees herself as more capable of taking care of the property than Pamela.

When we consider aggression as a function of territoriality, we see how aggression and hostility have come to be erroneously viewed as synonymous terms. Because establishing and protecting the territory is one of the strongest canine drives, we can understand that in this arena wild animals are quite willing to demonstrate a competitive edge. However, territoriality and its concurrent purpose, mate selection, primarily pits brother against brother; usually males of similar age and experience contend for the same territory or right to mate with the same female. They tend to be more equally matched and therefore back down less frequently than if the two were of different sex, age, or species. Under these circumstances, a fight becomes the only way to settle the conflict, and the expressions of aggression and hostility do look the same. While such apparent hostility is one form of aggressive expression, and certainly a dramatic one, we must remember it's not the only one.

For example, although we may assume that the winner of a fight has evidenced competitive superiority and won the battle for the territory, pack leadership, and the female, the winner could in fact lose the war. While he's been aggressively establishing and defending his territory and impregnating females, others may have been more aggressively seeking shelters or food. While the former passes on at least his abilities to fight to future generations, thereby perpetuating and hopefully improving the species, those aggressively involved in the less dramatic battles of daily life make an equally valuable contribution to survival. Were a situation to arise such as a severe storm or flood, in which secure shelter means more to survival than a willingness to fight or breed, those who aggressively pursued the former would be more likely to survive.

This more expanded view of aggression is expressed in the familiar phrase, "He who fights and runs away lives to fight another day." And remember the Wizard of Oz admonishing the cowardly lion to view his fleeing as a sign of his great wisdom, not his cowardice? Fear and aggression enable animals to achieve harmony with their surroundings in the most efficient way. While the willingness to do something or nothing at all produces the most stable relationships, it's the *un*willingness of one or both participants in a situation to do something or nothing or, worse, their vacillation between doing something and nothing that creates problems.

Therefore when fear and aggression create problems within a human-canine relationship, the real problem lies not in the manifestation of these states but in the relationship itself. Fear and aggression form a unique checks-and-balance system between the individual and its environment and between owner and pet. If the owner or the dog grows too fearful or too aggressive, the balance wavers, and the relationship crumbles. While problems involving minimal or intermittent fears or negative aggressive displays create imbalances similar to those encountered by mismatched submissive or dominant owners and pets and can be resolved using the same techniques, two situations deserve our special consideration.

FEARFUL OWNER, AGGRESSIVE DOG

How does the relationship between Pamela and Rocky following the biting incident differ from the submissive owner/dominant-dog combination we examined in the last chapter? The major difference is the presence and magnitude of fear. Prior to the Halloween biting incident Pamela and Rocky shared a much more passive relationship, in which Pamela accepted her subordinate position as well as Rocky's more dominant one, more or less willingly ceding him that position because she expected him to protect her, and above all, *never to hurt her*. But in the split second it took him to tear her arm open, her beliefs regarding his dominant, bossy, and protective nature gave way to one overriding feeling: "Rocky is a vicious dog who wants to hurt me, and I must be on guard at all times."

While we can't excuse Rocky's behavior, we now understand why it occurred. When Pamela and Terry put Rocky in charge of the household, Pamela's unwillingness and inability to control him were equivalent to her saying, "You're the boss, Rocky. I'll play by your rules." However, Pamela not only didn't know the rules, she didn't even know the game. While she thought they were playing "Protect Me from Burglars and Muggers," Rocky was playing "Protect My Territory from Anything *I* Consider a Threat." The games, their rules, and objectives, stood miles apart, so far apart in fact that when Rocky won his game, Pamela lost hers.

This little drama involves a canine actor who is overly eager to do something and a human actress who would prefer to do nothing at all. The only problem-solving steps the do-nothing person can take are either acceptance or getting rid of the dog. Pamela rejects these options because, although she'd like to blame the entire incident on the spaniel, Terry's work with the trainer and her new knowledge of pack behavior make that impossible. Furthermore, Pamela feels that her own level of self-confidence has sunk so low and her submission so great, it's highly unlikely Rocky can change enough to submit to her in her current state. Even if training could alter Rocky's behavior sufficiently to accommodate Pamela's existing self-image, she fears that the result might be less than ideal. It's possible that Rocky might only respond properly to her in Terry's presence; at worst, he might have to be reduced to a cowering blob to complement her current state. Because neither option offers the kind of relationship she wants with Rocky, Pamela decides to make some changes.

Because fear reinforces the do-nothing state, one way to eliminate fear is to do something. The more energy we channel into specific voluntary acts, the less energy is available for fear-based displays. We also know that the success of any action depends on purpose; the frightened animal that runs in circles urinating and defecating uses a great deal of energy but doesn't resolve the basic conflict. Similarly the owner who seeks to eliminate fear of the family pet by doing push-ups does little to resolve that inequitable relationship.

Because Pamela sees her problem as not fear itself but fear

of Rocky, she decides to make active changes that address her fear. Her first steps take surprisingly little effort, but they effectively boost her self-confidence. She invests in heavy leather boots and sturdy gauntlet gloves and dons denim jeans and jacket. When she wears this outfit, she feels secure; even if Rocky were to turn on her, her armor would protect her.

It's amazing how many owners won't follow Pamela's example because they don't want others to think them foolish or cowardly. Some even refuse to do it because they believe it reflects negatively on their pet: "I know Chipper doesn't mean to snap at me, so it's mean of me to insinuate that I don't trust him." Both excuses are pure hogwash and do nothing to solve the problem. The immediate problem isn't that Chipper bites or why he does so; the immediate problem is the owner's fear. To a dog a fearful owner looks the same as another fearful dog does: submissive. And the owner will look that way until he or she removes the fear. As long as fear intrudes, no owner can establish dominion over a dog.

The prospect of a fearful owner ever establishing dominion over a dominant aggressive dog seems unlikely, doesn't it? Yet, if such owners commit themselves to the relationship and deal with the fear *first*, the rest progresses fairly smoothly. Still wearing her heavy clothing, Pamela begins working with Rocky, first with Terry nearby, then alone. At first she tries to command Rocky exactly the way Terry does, but the trainer explains how that's neither necessary nor advisable: "You have to use a method that makes you feel comfortable, one that works for you. You're much better off using one unlike Terry's than a weaker, diluted version of his." He also urges her to practice giving authoritative commands, replacing her apologetic sing-song whine with slightly louder, deeper, and more precise tones.

As she practices her skills and works with the dog, her confidence grows. Like most owners involved in similar relationships, however, she eventually reaches a plateau. Rocky responds well to her, seems to enjoy her handling, and makes no hostile moves after the fateful Halloween; on the other hand, he hadn't made any before then, either. That thought bobs persistently in Pamela's mind for over two months as she dresses for Rocky's daily workouts.

The willingness of dog or owner or both to change makes the relationship between dogs and owners very special. When emotions as devastating as fear and hostility undermine the relationship, however, our lack of confidence sometimes persists along with the memory of the upsetting incident. Whenever fear acts as a barrier, we can find it so insurmountable that we want a sign, an event as strongly positive as the negative one was negative, to neutralize the bad memory.

When such a sign comes (and it always comes to those who want it), it carries its own special healing magic. As part of Pamela's program of interaction with Rocky, she grooms him daily. One day during a romp in the woods the exuberant spaniel plows through some dense underbrush and emerges with dozens of burrs entangled in his silky coat. At first Pamela works at them gingerly while Rocky lies contentedly on the floor. As she becomes more engrossed in her task, she pays more attention to the burrs and less to the dog, even when she attempts to remove the burrs from the highly sensitive areas around his rectum and penis. As she tugs too hard, Rocky tries to pull away, but she ignores him and pulls harder, obviously increasing his discomfort, until Rocky spins around growling and presses his nose hard against her hand.

The gesture so startles Pamela that she freezes and feels tears well up. Her mind whisks through assorted emotions: fear, anger, fear, frustration, fear, fear . . . But Rocky doesn't bite; he doesn't run. He just sits there looking at her and pressing his cold, wet nose against her wrist. Something clicks in Pamela's mind and makes her stop emoting and start thinking. In Rocky's eyes she reads, "I could have bitten you, *but I didn't.*" Until Rocky exhibited the former display but stopped short of the bite, Pamela wanted to believe, but had no proof, that Rocky didn't want to hurt her.

While it would be nice if our faith in our pets made proof of their devotion and trustworthiness unnecessary, anyone who's experienced dog-based fears of any kind knows that's easier said than done. The important thing is to muster the confidence in ourselves and our dogs to know that proof will surface when we need it most.

For Pamela the awareness that Rocky could have bitten her but didn't, even though he had good reason to do so, removed the final obstacle standing between her and a solid relationship with her pet. By choosing to trust her dog and herself and wipe away the remainder of her fears, Pamela is finally ready to accept dominion over her dog.

AGGRESSIVE OWNER, FEARFUL DOG

Just as fearful owners of aggressive dogs may find themselves with frightful problems on their hands, aggressive owners of fearful dogs face just as tough a time creating a good relationship. Surely no one could imagine that poor Kimo, a member of one of the most heroic working breeds, enjoys scurrying under the bed every time he hears a loud noise. However, we can see how thoughts of the Akita's noble heritage could lead Howie Chan to conclude that he owns a canine dud. The fearful-dog/aggressive-owner is a surprisingly common combination, and in the typical situation the owner actually feels threatened by the dog's fear, yet knowingly or unknowingly perpetuates it. This typically occurs when owners expect the dog to be fearless (in order to protect them) and/or see the dog as an extension of themselves. Howie's view of Kimo's proper behavior includes both concepts. The Akita's heritage indicates that the dog should carry himself bravely; and Howie doesn't want to be known as the owner of a cowardly dog.

Kimo considers thunder life-threatening, the regrettable consequence of a cherry-bomb attack by the Chan twins when he was in the fear-imprint stage. Because he can't see thunder, he can't fight it; so he chooses to run as far and as fast as he can—which usually means to the farthest corner of the master bedroom, under the bed. When Howie blocks Kimo's chosen path for dealing with the threat, Kimo can either freeze or fight. However, he recognizes Howie (and all the Chans) as dominant, an orientation the family has reinforced since Kimo was very young. So as the yelling Howie looms before him, Kimo's life now contains two fear-producing demons:

- The thunder from which he wants to flee.
- Howie, in whose presence he normally freezes.

A second loud clap of thunder convinces Kimo he should run rather than freeze; thunder is life-threatening, Howie isn't. However, Howie takes the dog's fear so personally that he reacts, or rather overreacts, to counter it. Ostensibly he wants Kimo to pay more attention to him than to the thunder so that he can allay his dog's fears. In reality he wants Kimo to respond to him to the *exclusion* of everything else, even something the dog believes threatens his very life.

Howie throws the toy truck with the idea of distracting the frantic dog with a louder, more immediate noise than the thunder. In theory that tactic might work in a more stable relationship, but Howie's aggressive orientation toward his dog usually creates threats and challenges for the pup to endure. The appearance of this third threat simply panics the inexperienced Akita, who emotionally and physiologically loses it. Because he doesn't know what to do with Howie blocking his chosen response, and no appropriate alternatives are in sight, he tries to display everything in hope that one of his signals will do the trick. He winds up displaying fear and aggression—urinating, defecating, and expressing his anal glands involuntarily while racing frantically around the room.

In cases like this the last thing the dog or the situation needs is an aggressive owner shouting and charging around too. While Howie certainly didn't cause Kimo's response to the thunder, he certainly did nothing to lessen it and actually made it worse. The wise Connie takes the problem in hand by shooing Howie out of the house, letting Kimo crawl under the bed, and cleaning up the mess.

When Howie calms down, the family agrees that they want to keep Kimo and help him, and the Chan boys start the ball rolling in the right direction by confessing to the cherry-bomb caper. Rather than being angry, their father actually feels relief:

Now he can acknowledge a *real* reason for Kimo's fear and a far better one than his belief that his dog suffers from inherent cowardliness that he, as an incompetent owner, has been unable to overcome. While the Chans could initiate a training program whereby they gradually accustom Kimo to ever louder recorded sounds of thunder and thus diminish the Akita's fears, what about Howie's excessively forceful relationship with Kimo?

Howie dismisses the incident easily, "Oh, I only got mad because he was such a coward." His family and friends doubt his words. At the center of the Chans' problems sits the relationship between an aggressive owner and a fearful dog, not the one between a frightened dog and thunder. With patience and consistent training we can extinguish Kimo's fears, but unless Howie relaxes a bit, he'll find another behavior for which to chastise and criticize the dog. For example, let's assume that Kimo gets over his fear of thunder about the time he becomes sexually mature and starts marking his territory. When Howie spots the dog lifting his leg on a shrub, he leaps at him, shouting and waving his arms threateningly. Far from being a relatively harmless show of dominance, Howie's gestures are excessive, unnecessary, and inappropriate. Kimo already recognizes Howie's dominance, so a simple "No" would suffice; the reaction is inappropriate because Howie wants Kimo instantly to redefine and modify an instinctive biological activity, interpret it as "wrong," and freeze submissively with one leg midair, a physically and behaviorally impossible feat.

Surprisingly, while aggressive owners often consider themselves supremely confident, in fact the opposite is usually the case. *Lack* of confidence leads people to overreact. The truly dominant, masterful individual in any situation need only signal its (or his or her) willingness to act and will actually act only when the other doesn't recognize his, her, or its authority. Owners who overreact should reevaluate the purpose of their responses and body-language displays toward their dogs. It could be that they, too, are playing a different game with different rules than their dogs: Only this time the dog gets hurt.

Unconfident people who get dogs to prove their power via their ability to precipitate fearful or hostile canine displays can only maintain their position by keeping their dogs in an unnatural state. The "respectful" freeze or "protective" fight displays aren't meant to be part of the dog's daily routine; rather, they're part of an intricate system of self-preservation. When Howie expects Kimo to freeze in the middle of urination or to ignore his strongest instincts to flee from thunder just to prove Howie's authority over him, the relationship is aberrant to the point of being grotesque. If owners like Howie can't resolve their own confidence problems with or without professional help, they should terminate the relationship for everyone's sake. Like people, dogs shouldn't continually live in fear; their entire being revolts against such an unnatural state. When dog and owner share the fear but manifest it in antagonistic ways, a bad relationship will invariably become terminal one way or another.

THE BONDED VIEW OF FEAR AND AGGRESSION

Bonded owners recognize fear and aggression as components of an extraordinary protective mechanism available to the threatened animal, not as a way of life. Together they work like the emergency brake and first gear on my Subaru. The brake keeps me from moving when it's safer to be still; first gear provides a lot of power to get me out of a tight situation quickly. And although each device performs its function, neither serves any purpose while cruising full-speed down a highway. Imagine trying to slow down by applying the emergency brake while traveling at fifty miles per hour, or attempting to maintain that speed in first gear. Both mechanisms fail miserably when applied inappropriately and cause tremendous and unnecessary wear and tear on the whole system. The same holds true for fear and aggression; they work well in truly threatening or demanding situations, but severely retard the growth and development of a relationship if they become continuous motivating forces.

The normal day-to-day functions of a relationship are best accomplished via the interplay of dominance and submission, with a constant appreciation that we do some things better than our dogs, and they do some things better than us. Any sound relationship springs from inter*play:* No owner needs to create or maintain an extreme state of submission or fear, nor dominate or be dominated to the point that competition, intimidation, and negative aggressive displays taint the relationship.

As always, self-confidence is the key ingredient in the interplay of dominance and submission without fear and aggression. If we lack confidence in ourselves and our pets, we can fall into the burdensome trap of thinking that their constant submission and fear prove that they need us or into the equally dangerous trap of thinking that their dominance and aggression prove that we need them. Either relationship based on fixed roles and mutual needs inevitably creates problems because it arises from a perceived deficiency in one or both of the partners.

By contrast, the bonded relationship, based on the belief that we live with our dogs because that's what we and they *want*, builds on strength rather than on weakness. If we want to be together, the relationship at any given moment means more than who's playing what role. Because we recognize that we want to be with our dogs and *vice versa*, accepting dominion over them isn't a burden; it's an unemotional fact of life. Finding creative ways to exercise dominion over the fearful or negatively aggressive animal may in fact provide us with some of the most challenging and rewarding emotional experiences dog ownership has to offer. Because we want to be with our dogs, to bond with them and share experiences with them, we banish fear and competitiveness from the bond. What possible benefit can we gain from being afraid of our best friends or vice versa? Can a competitive relationship bring any great joy, even to the apparent winner? How much confidence do you derive from believing you depend on or need your dog for protection? How much confidence does it give you knowing your dog depends on you for its well-being, that without you it would die of physical, mental, or emotional starvation?

Bonded owners recognize that the ideal dog could live with anyone, but chooses to live with them. They relate to and train their dogs to reflect this belief and consequently shun cultivating submissive, dominant, fearful, or aggressive displays in themselves or in their dogs that would preclude interactions with others.

The bonded relationship between human and dog forms a

fulcrum balancing dominance and submission and their extreme protective forms, aggression and fear. As long as the relationship remains stable, the balance may tip in one direction or another and, if necessary, to the extreme limits but will quickly stabilize. On the other hand, fixed relationships are static and provide little opportunity to experience the full range of one's own abilities and those the other has to offer. Because of this, such fixed orientations ultimately result in relationships plagued by boredom and resentment. Compare this to a bonded relationship based on confidence and a willingness to accept dominion over the dog; it can survive all kinds of challenges and still remain stable.

The power of the bond often escapes us because it runs so deeply through our relationship that we're not even aware of it. But when our bonded dogs take a cue from our anger or fear and heroically rise to the occasion, we see it in full bloom. Read the articles about courageous dogs: Nine times out of ten the owners express shock that their perfectly normal happy-go-lucky pets were capable of such heroism.

A married couple strongly bonded with their two golden retrievers provided irrefutable and poignant evidence of this phenomenon. When the dogs were four and six years old, the marriage began to falter. At first the couple argued, then they began throwing things at each other. The dogs' normally seasonal allergic skin problems persisted; they lapped and chewed themselves much more and didn't respond to medication as they had in the past. Their once-beautiful silky coats turned brittle and dry from lapping; some areas of skin were completely denuded, and oozing sores marred once-flawless skin. Worse, the dogs paced and whined constantly, wearing the most confused and distraught facial expressions I've ever seen on any animal. I was shocked by their appearance.

"Good Lord, what's bothering these dogs?" As the wife told me about their troubles, I could picture husband and wife each giving cues that would have triggered the dogs to attack had they come from anyone else; but since both "attackers" were beloved owners, the dogs took out their frustration by lapping themselves and each other frantically, pulling out great wads of fur. When I saw the dogs twelve hours after the husband stormed out of the house following what was to be the last violent confrontation between the couple, the goldens had yet to eat, drink, or stop pacing and whining.

It took months for those dogs totally to recover from the mental, emotional, and physical trauma that resulted from the shattered bond. While definitely a tragedy, it left me with a sense of awe regarding the power of the bond and the delicate balance of its components.

Compare the energy of a dog that romps joyously with the kids and interacts constantly with its family and friends and unfamiliar people and events to that of the one-man dog who perceives practically everything as a potential threat. Which animal uses its physical, mental, and emotional capacities to the fullest? Which owner has forged a strong bond on *all* levels?

Truly bonded owners not only have confidence in themselves, they also have confidence in their dogs. Recognizing that the normal nonthreatening times they spend with their pets far exceed those when they, their family, and friends are threatened, they develop their relationships based on the ordinary; in the process, they create a relationship that will survive even the most extraordinary challenges.

This ends our discussion of dominance, submission, fear, and aggression—the primary states from which all other behaviors and emotions spring. Now that we recognize what a critical role relationships play in the survival and stability of the individual, it should come as no surprise that severing that relationship, even for only a few hours, can create strong body-language displays and emotional repercussions. We touched on the implications of isolation when we discussed the fate of the highly dependent pet. Now let's take a look at that common modern canine phenomenon in depth. What happens when we isolate a social animal like the dog? Should we feel guilty when we leave Wookie at home? Is it normal and understandable that he destroys the house while we're gone?

8

BOREDOM, FRUSTRATION, AND ISOLATION: THE TERRIBLE TRIO

SONJA JOHANSSEN DASHED into the crowded subway train and miraculously found a seat. Then, sitting with her briefcase jammed between her legs, her double-strapped pickpocket-proof handbag clutched on her lap, she regretted having moved to the city. How she longed to pull out her wallet and look at the snapshot of her parents with their Great Pyrenees, Snickers and Boris, in front of the barn on their Wisconsin farm! But Sonja worried about opening her purse on the crowded train. Oh well, next week's paycheck from the law firm where she worked would probably make it all seem worthwhile again. Maybe she'd treat herself to dinner out or a nice steak.

Food. Darn, she'd forgotten to buy dog food again. The thought of her six-month-old male basset hound, Bernie, simultaneously cheered and alarmed her. During the four months Sonja and her dog had lived together, Bernie had provided comfort and companionship in a seemingly impersonal and even hostile world, but at the same time he had, in her absence, systematically attacked virtually every accessible object in her apartment. Five days a week Sonja devoted at least half of the commute to and from her midtown office agonizing over her relationship with Bernie:

- What an attentive and marvelous companion he was!
- How could he so ruthlessly destroy her favorite plants?
- It's cruel and inhumane to own a dog in the city; she should find a suburban home for Bernie.
- What a joy to have such a natural creature in this artificial environment!

As if matching the rocking rhythm of the train, her thoughts swang from "He loves me" to "He loves me not."

When Sonja finally enters her tiny apartment, Bernie flies into her arms yapping joyfully, his entire body wriggling ecstatically. As usual, Sonja's heart quickens at this display of affection, then chills at the sight of the shredded aspidistra, the strewn trash, and the gnawed leg of her grandmother's armoire.

Whether or not we ever take time to consider the roles boredom, frustration, and isolation play in our lives, they are nevertheless three of the most troublesome emotional states we and our dogs experience. Consider two of my favorite descriptions of these emotions and their effects on humans: (a) the "nyah-nyahs," which means "nothing will please me, so don't even try"; and (b) "When I'm frustrated, bored, and alone, I feel like someone ripped my head off, filled my body with sand, and screwed my head back on upside down." These may not constitute any particularly objective accounts of boredom, frustration, and isolation, but they do nicely capture the ways these emotions make us feel. And given our close associations with our pets, we can be pretty sure that what bothers us will affect them, too, in one way or another.

THE TWO-HORNED DILEMMA

In this chapter we're going to consider the body-language expressions, emotions, interpretations, and responses associated with boredom, frustration, and isolation together because, in reality, we can't productively discuss one without winding up discussing the others. Unresolved boredom leads to frustration; frustrated animals exhibit negative behaviors that often lead their owners to isolate them. Nondestructive animals who are isolated may become bored and frustrated, then destructive.

Boredom, frustration, and isolation present owners with a two-horned dilemma: First, we must confront any actual damage associated with the dog's expression of these emotional and behavioral states, then we must deal with the emotional trauma we suffer whenever we believe our dogs have fallen prey to these emotions.

For example, when Sonja unlocks her apartment door after leaving Bernie alone all day, she recognizes and responds to a familiar sequence:

- Bernie's body-language expressions.
- Her interpretation of the underlying canine emotions that led to those expressions.
- Her emotional response to what her dog has done.
- Her own body language reflecting her emotions.

During a typical homecoming, Sonja first sees a deliriously happy dog, then the destruction. She usually interprets the former to mean that Bernie loves her, the latter than he took out his boredom and frustration on her by chewing the plant, scattering the trash, or gnawing the furniture. Depending on her own mood, these conflicting body-language signals and emotions may cancel each other out, enabling Sonja to remain calm, petting Bernie distractedly and cleaning up the mess while thinking about her job or something else entirely unrelated. If Bernie's cheerfulness exceeds the destruction, she may respond positively to him—"Oh, Bernie, how wonderful, you only ate half a plant today!" If she's depressed or had a bad day at the office, she may reward the basset with a hard smack for the same body language and emotional display: "Bernie, no! Bad dog!"

Compare Sonja's plight to that of retired Brigadier General A. J. Matheson, his wife, Carla, and their Staffordshire terrier, Fezziwig, who live in a rambling Croton-on-Hudson split level. The Mathesons are suffering from the combined affects of retirement and an empty nest. A.J. has been out of the military over a year, and preparations for their only daughter's gala wedding have filled the Mathesons' days up until about a month ago.

"At last we can start relaxing and enjoying ourselves," A.J. told his wife. "You can finally take that course in flower arranging at the college and, by golly, I'm going to knock six strokes off my handicap."

Within a month, however, Carla dropped her course, A.J. stopped playing golf, and the two of them rarely went out. "We just can't bear the look on Fezzi's face when we leave him alone," Carla wrote their daughter a short time later.

In this case the dog has shown no negative physical signs of boredom, frustration, or isolation, no soiling, chewing, or unsightly sores or raw feet to indicate he's been lapping himself during the owners' absence: only that telltale "look." As A.J. describes it, "He goes real still and then kind of looks right through you, almost as if you're not even there. It breaks Carla's heart and makes me feel like I've just court-martialed Rin-Tin-Tin!" Whereas Sonja can provide a long list of body-language evidence indicating Bernie's boredom, frustration, and isolation, the Mathesons can cite only Fezziwig's vaguely defined "look." However, even though we may be tempted to consider Sonja's relationship with her dog in more jeopardy because of the destructive nature of the behavior associated with it, the Mathesons' problem is equally burdensome to them.

ACTIVE AND PASSIVE BODY LANGUAGE

These two examples illustrate active and passive body-language expressions of our three emotional states. Among the common active expressions we may include:

- Chewing.
- Digging.
- Scratching.
- Vocalizing (barking, whining, howling).
- Soiling.

These expressions manifest according to the circumstances, canine temperament, and even breed. Some dogs may sleep contentedly if left alone in a car, but soil and bark frantically when isolated in a small room. More introverted canines tend to chew, dig and scratch at themselves, whine and urinate submissively, whereas more extroverted and dominant ones chew and scratch objects, dig holes, bark, and mark their territories with urine and stool. Mouthier bird-dog breeds (setters, retrievers) will more likely lick and chew; more vocal scent hounds (beagles, bassets) will usually bark and howl, and terriers will try to dig out or in.

The active body language of boredom, frustration, and isolation comprises a virtual rogue's gallery of negative canine behavior, so it comes as no surprise that some behaviorists attribute 90 percent of all the negative canine behavior to these states. Like humans, dogs have evolved in the company of others; when they're denied that companionship and something equally attractive, stimulating, or comforting to replace it, their self-confidence and stability deteriorate. Any insecure animal wants to regain its security. Humans may smoke, drink, go on eating binges, or chew their nails; dogs chew, dig, and scratch themselves or anything they can get their claws on.

On the other hand, dogs who engage in passive body-language expressions don't cause physical destruction. Their owners rely on less obvious cues such as sighs, listlessness, and dull, sad, or hurt looks. Unlike the chewed table leg or oozing sore, these indicators are highly subjective. To perceive them, owners must either be highly attentive and sensitive to their pet's most subtle body language or develop very strong beliefs about what their dog is thinking and feeling. Often it depends more on what the dog *isn't* doing than what it is. If Carla expects Fezziwig to wag his tail joyfully when she leaves the house or rush to meet her when she returns but her pet does neither, this lack of expression means as much to her as a mangled rug does to Sonja. And although Sonja can point to concrete evidence of Bernie's emotional state, while the Mathesons have only Fezziwig's ill-defined body-language expressions, the Mathesons' problem could actually become the more severe.

Because Sonja associates destruction with Bernie's misbehavior, she always knows when the basset does something wrong; if she sees no damage, she perceives no problem in her relationship with him. In contrast, the Mathesons' lack of concrete evidence leads them to equate a vast array of subtle signs with Fezziwig's "misbehavior." Bernie may get scolded or smacked every other day, but Fezziwig's owners could lavish treats and affection on their pet one day to assuage these subtle signs and punish or alienate him for the same expressions the next. Whereas Sonja experiences some homecomings that aren't marred by Bernie's misbehavior, the Mathesons believe Fezziwig is *always* unhappy about being left alone.

THE CANINE EMOTIONAL BASE

Because we're dealing with three different yet related states, we naturally see a broad range of body-language expressions. Surprisingly, owners seldom attribute such behaviors to these states. Even when they do associate the behavior with boring, frustrating, or isolated environments, they tend to attribute it to the dog's spitefulness, meanness, stupidity, love, or devotion. Whereas the behaviorists would say that Bernie chews Sonja's favorite book because he's bored, frustrated, or isolated, Sonja believes he does it to get even with her for leaving him alone. Owners like the Mathesons who recognize passive displays in their pet may also attach such feelings to their dog's behaviors, interpreting certain postures as "I'm sad because *you* left me alone," rather than "I'm sad because I'm alone."

In either case owners begin to perceive themselves as the target or object of the dog's emotion and subsequent body language. The question is: Is this a valid supposition for us to make? Obviously when Sonja leaves Bernie alone all day, she can correctly define herself as the cause of his isolation; if she stayed home, he wouldn't be isolated, bored, or frustrated. On the other hand, we also know that the dog's retention span, its ability to associate one event with a later one, is quite short— some say as little as thirty to sixty seconds. Therefore, during this brief period immediately following Sonja's departure, Bernie may associate any loneliness, boredom, or frustration with his mistress's departure, but he certainly won't do so an hour later when he rifles the garbage. So how does Sonja arrive at this conclusion? Like this:

> "I know Bernie really loves being with me, so he must be terribly upset when I leave him alone. Once I leave, I'm sure he tries to figure out what he can wreck that will hurt me the most."

What about those days when Bernie doesn't touch anything of importance?

> "I think that happens after we have an especially good time together or when he knows I'm going to be home more. Also, if I get really upset about something he ruined or really mad at him, he sometimes feels guilty the next day and behaves himself."

Notice how Sonja not only credits her dog with a retention span lasting days, she also believes him capable of making some very complex human associations. Whereas the behaviorists would describe Bernie as a social animal who doesn't like being alone or in a dull environment and therefore tries to dig, scratch, or chew his way out, Sonja believes the basset evaluates their previous interactions, reflects on them, and decides whether or not to punish his mistress. If Sonja agrees with Bernie's evaluation—"No wonder he chewed the bedroom door, I've hardly been home all week!"—she considers the chewing an expression of his justifiable anger. If she disagrees—"Doggone it, Bernie, I spent all weekend with you and look how you repay me!"—she detects spitefulness or even stupidity in his behavior: "How many times have I told you to leave my shoes alone, you dumb mutt!"

Meanwhile the Mathesons make a completely different connection. When A.J. reflects on Fezziwig's apparent sadness he notes:

"Old Fezzi's been with us for almost fifteen years, and when our daughter lived here, someone was always around. Now that the house is so quiet, the old guy can't stand it."

While the behaviorist may willingly attribute Bernie's destructive activities to the unemotional states of frustration, boredom, and isolation, chances are they'd consider Fezziwig a perfectly normal, well-behaved dog. If he weren't, if he experienced any boredom or frustration when left alone, he would vent his feelings on the environment or on himself.

"That's not true," interrupts Carla. "Fezzi's like A.J. He keeps things bottled up inside. Also, he's so well-behaved, he'd never destroy anything, and so clean, I can't imagine him ever chewing himself. That's why we have to watch him so closely to figure out what's bothering him."

To our array of active and passive body-language displays associated with boredom, frustration, and isolation, we may add an equally long list of canine emotions owners believe precipitate those responses. Balancing this, we have the behaviorists' view that dogs resist limited and/or isolated environments because they're curious, social animals. When these basic needs are thwarted, the dog attempts to achieve a more favorable state, either by gaining freedom from the environment or relieving the associated stress.

THE HUMAN EMOTIONAL RESPONSE

Having acknowledged the canine body language and assigned emotions to it, we must now respond to our own creation.

"Hey, *I* didn't create that mess in my apartment," retorts Sonja. "Bernie did!"

That's true, but we're not talking about the mess; we're talking about Sonja's belief that the mess results from her dog's anger, spitefulness, or stupidity. Having made such an association, how does she feel about the result?

"When he doesn't misbehave, I'm so relieved I hug and kiss him and tell him he's wonderful. If he's done something wrong, it depends. If he chews a sweater I left lying on the kitchen floor, that's my fault; but if he pulled it off the hanger in my closet, that's his fault. Also, if I took him for an exceptionally long walk or gave him a special treat, then the slightest misbehavior really bothers me. Sure, my own moods are a factor, too—sometimes I'm just too tired to care, sometimes I'm really furious for no good reason, and still others I'm so tolerant and kind I even surprise myself. It's terribly confusing."

When we analyze all the different isolation behaviors Bernie exhibits and the various canine emotions Sonja attributes to them, we can quickly become confused ourselves. Rather than clarifying the relationship, the addition of Sonja's emotional responses now muddies the waters even more. In addition to a wide range of canine body-language expressions and canine and owner emotions, we also have a relationship in which the identical body language can be linked with a wide variety of human and/or canine emotions on any given day. Even as a professional who regularly copes with canine behavioral problems, I find juggling all this conflicting data difficult. No wonder Sonja's exhausted!

Although Sonja might think she'd love to trade problems with the Mathesons, A.J. and Carla agonize over their deteriorating relationship with Fezziwig:

"We've been looking forward to retirement for years," laments A.J. "Now here we are trapped at home by a dog. We love Fezzi, he's a grand old fellow and was so devoted to Carla and our daughter all those times I was gone. I feel guilty as hell about leaving him alone, but I resent staying home because of him, too. I've started toying with the idea of having him put down—and that really makes me hate myself."

So even though Fezziwig does nothing more than heave a few barely audible sighs and muster a sad look or two, the "lonely," terrier elicits a virtual Pandora's box of negative emotional responses from the Mathesons, including sadness, guilt, anger, and resentment. And even though the behaviorist may argue that Fezziwig isn't displaying any true frustration- or isolation-based behavior whatsoever, the fact that the Mathesons believe he is makes his "misbehavior" quite real to them.

The ways in which others interpret our dog's behavior also contribute to the problems associated with loneliness. Because active isolation behavior tends to be destructive or irritating, owners of dogs exhibiting it often find themselves at odds with neighbors, friends, and even the law. As a working owner, I once came home to find a warning from the dog officer taped to my door; someone had complained about my dog's barking in my absence. Although I first experienced shock, embarrassment, anger, and frustration, all aimed at the unknown complainer, myself, and my dog, I finally determined to rectify the problem. On another occasion, some friends laughed but others shook their heads in disbelief over the elaborate barricades I'd constructed to keep my dog off the couch when I was gone. In this case, I felt comfortable with my jerry-built solution, so I was immune to their remarks.

If we can't accept any adverse comments, *we* wind up feeling isolated, too, particularly when neighbors complain to the law or others rather than directly to us. Although they may rationalize this by saying they don't want to hurt our feelings, their indirect criticism hurts even more. Imagine dining in a restaurant with fifty other patrons when the maître d' leans over and says, "We've had a complaint about the way you're eating your soup. If you don't stop, you'll have to eat in the kitchen." Perhaps you're eating your soup the same way you have for years and no one ever complained before. Or perhaps you realize you do slurp your soup and want to change but don't know how. Regardless, you suddenly feel like a criminal surrounded by a roomful of accusing eyes.

Similarly, most owners who receive official warnings, summonses, or anonymous phone calls feel like criminals. Suddenly we and our dogs have sinned against society, our previously friendly neighborhoods have become hostile territory, and worse, our homes can become prisons. In our hurt, anger, and depression, we lock ourselves and our dogs away from our critics. When we're home, we severely discipline our pets, and whenever we must leave them, we take extraordinary measures to ensure that they can't possibly bother anyone. The whole time we're gone, our thoughts leap home time and time again: Is Rory barking? Are the Smithfields home? Please, God, don't let the McDougals be working outdoors today!

Before long we find ourselves buckling under a tremendous burden of negative emotion, but still lacking viable solutions to our problems. When people complain to me about how other people's dogs behave, I invariably recall one of my favorite jokes: Two cars are stopped at a red light. A woman sits behind the wheel of the first car, a subcompact bulging with several young children and bags of groceries. Behind her, an impeccably dressed businessman commands the helm of a fully appointed luxury sedan. The instant the light changes, the businessman honks his horn. In her surprise and haste to move, the woman stalls her car. The businessman continues blasting his horn, so flustering the woman that she floods the engine. The man keeps blowing his horn, the kids begin to cry. Finally the exasperated woman climbs out of her car and walks back to the obnoxious driver: "Look, why don't I sit back here and blow your horn while you go up and start my car?" It's a lot easier to make a lot of noise than it is to solve a problem.

As hard as it may be to do so, it pays to be as objective as possible when dealing with other people's point of view. If some anonymous caller insinuates that you're a rotten owner because your dog barks all day, eliminate your emotions and examine the facts: Does Rory bark all the time? Most of the time? Not at all? Quite possibly, the barking occurs only when you're gone; then you'd never hear it and could therefore be tempted to call the complainer a dog-hating jerk. Before you discard your positive attitude toward your neighbors, make a few inquiries. Forget your bruised ego, anger, and other bad feelings; make an information-collecting tour of the neighborhood. Be open and honest: "I received a warning about Rory's constant barking, and I'm not sure what to do. He rarely barks when we're home, but I have no idea what he does when we're gone. Have you ever heard him?" Be sure to ask "Have you ever heard him?" rather than "Has he ever bothered you?" The former question elicits factual information; the latter demands a subjective judgment others may be reluctant to make. In such a way you may realize that your neighbors do have a legitimate complaint. Then you can objectively analyze the behavior and maybe even engage your neighbors' help in solving the problem.

CONVERTING HUMAN EMOTION
TO BODY LANGUAGE

How do owners manifest their emotions: What body language do we use to convey our emotional evaluations of our dog's bored, frustrated, and/or isolation behavior? Sonja frowns. "I laugh, I cry, I yell. I throw things—sometimes at Bernie, sometimes at the wall. I sulk, I ignore him. I eat a whole lot or drink a bottle of wine to get my mind off what he's done. I smack him with my hand, a rolled-up newspaper, once I hit him with a loaf of French bread." Just as Bernie's body language elicits any one of an infinite variety of human emotional responses, so Sonja manifests these in any one of an infinite number of body-language displays. To ensure the consistency it takes to make training effective, Sonja would need a computer!

As we would expect, the Mathesons don't fare much better: "When I see that look on Fezzi's face, I get so upset, I sometimes cancel my engagement and stay home with him instead," says Carla. "Other times I get angry and almost run out the door, and try to forget him while I'm gone; but when I come home, I have to face him again. If I feel really bad, I'll spend time petting and hugging him and will often stop at the store to get him an extra special treat. But if I had a good time with my friends, I may shout at him or ignore him because he's ruining it for me."

"I feel pretty much like Carla," adds A.J. "When it got to the point I was sneaking out of my own house to avoid seeing

him or going miles out of my way to buy him a T-bone, I decided it would be easier to stay home. However, if I don't attend some important function, I'm usually gruff and impatient
with Fezzi." On such occasions A.J. may allow his dog only a
few minutes to relieve himself, rush through their normally lei-
surely walks, or ignore Fezzi's invitations to play and share affection.

Although Fezziwig's subtle body language might be imper-
ceptible to everybody else, the Mathesons' interpretations of
the emotions behind it lead them into complex body-language
displays. In their case, Fezzi's expression almost always strikes
them as sad and lonely. Whereas Sonja and Bernie frequently
share a positive greeting, provided the basset hasn't destroyed
anything or that Sonja chooses to ignore it if he did, the Mathesons always judge themselves harshly and respond accordingly
whenever Fezziwig gives them "that look."

DEALING WITH OUR INTERPRETATIONS

Having analyzed two quite different situations involving
canine and human emotion and body language associated with
boredom, frustration, and isolation, we're ready to test them
against our familiar four options. Can Sonja and the Mathesons
accept their interpretations just as they are, including the effects on their relationship with their pet? Sonja, A.J., and Carla
ring forth with a resounding "No."

However, some owners may not share their opinion. When
Dick Petowski accepted a high-paying job in Chicago and
moved himself and his Siberian husky, Shane, from a remote
cabin in Minnesota to a crowded condominium complex, he
told himself that both he and the dog could cope with a tem-
porary but necessary evil. "It was too good an opportunity to
pass up. In three years I can make enough to start my own business and then head back to the woods for good." Because Dick
hates the city, he interprets Shane's isolation behavior as his
pet's way of saying he hates the city, too. Although Dick doesn't
like having his belongings destroyed by his dog, Shane means

the world to him. When Dick encounters the nightly mess, he apologizes to his dog: "I know, old buddy, I know just how you feel. But hey, I'm working as hard and fast as I can to get us out of here."

By preserving his view of Shane as a semiwild creature who deserves to be free, Dick uses his dog's negative behavior as an incentive to work hard and make more money. Were his canine best friend readily to adapt to big-city living, Dick might question his own reluctance to do likewise; if he adapted as smoothly, maybe they'd never get back to the woods. Because Dick dreams of returning to the country and believes the city is a bad environment for him and his dog, he'll interpret his dog's body language in terms of those beliefs. And even though Dick was upset by Shane's chewing and disciplined him for it when they lived in the cabin, he now accepts it as valid behavior for a wolf-dog trapped in the city.

Rose Lively also accepts destructive behavior from her Norwich terrier, Nova, and the negative responses it precipitates—but for entirely different reasons. Rose spent the last three years literally enslaved to her terminally ill husband. During their last year together, Ralph Lively's physical condition so deteriorated that he required constant care; concurrent mental degeneration caused the invalid to abuse the woman he'd adored for over thirty years. But Rose cheerfully stuck it out, recalling numerous fond memories to ease her through the difficult times, relying on Nova for comfort and companionship during those long days and nights.

When Ralph Lively died six months ago, everyone agreed it was a blessing. Finally all the physical and emotional pressure fell away, freeing Rose to renew old friendships and reassemble her life. Nova, however, didn't fare nearly so well. After being accustomed to constant company for years, the terrier exhibited classic isolation behavior. Although Rose doesn't like having her house disrupted and feels deeply hurt by Nova's misbehavior, she accepts his response as justifiable. Furthermore, she also feels somewhat guilty because she's enjoying her new life while Nova is evidently so miserable. Accepting Nova's behavior as understandable and believing that she deserves some sort of retribution for accepting widowhood so painlessly, Rose willingly accepts her dog's behavior as well as her own negative emotional and body-language responses to it.

Our second option involves accepting the canine emotion and body language but changing the way we feel about it. Sonja also rejects this option; although her furnishings may not be heirlooms, she doesn't want them destroyed. Furthermore, even if she could learn to accept the behavior and her feelings about it right now, she anticipates changes in her life-style that may place even more demands on her relationship with Bernie: "This might be the most time we have together. If I accept his behavior and convince myself not to let it bother me, I may live to regret it—especially if I buy new furniture or move into a better apartment."

The Mathesons also reject this option: "Even if we could convince ourselves that it's alright for us to feel angry and guilty about leaving Fezzi at home, I don't see how it can last," comments A.J. "I've been in the military long enough to recognize a mind game when I see one. Fezzi's been too much a part of this family for me to make believe that what he does doesn't bother me when I know it does."

However, a year after her husband's death, Rose Lively does select this option: "I used to accept Nova's destructive behavior because I felt he really missed Ralph and because it filled a need in me. I was used to cleaning up after others—that had been my whole life for years. And I know this may sound strange, but when Ralph died, I didn't know what to feel: How could I possibly feel relief? Nova stabilized me during those months, giving me the punishment I believed I deserved for feeling relieved. But that's all in the past now."

As Rose's life settles into a routine and her self-confidence returns, she begins to question the idea that Nova is punishing her. She contacts a canine trainer-counselor, who assures her that Nova has been responding to his environment, not her. Although he tells Rose that she can alter the Norwich terrier's behavior, Rose decides she can live with an upset wastebasket and a few shredded newspapers without feeling ill will toward her dog. Because Rose doesn't mind cleaning up (and blushingly admits she sometimes even enjoys it), she chooses to see the body language in a new light. Now when friends ask her why Nova tips the trash, she confidently tells them he's a social animal expressing his frustration at being left alone. "You mean he's getting back at you for going out?" they ask. "No," corrects Rose, patting her devoted companion, "That's just his way of relieving his tension."

How do Sonja and the Mathesons feel about the third option, changing their own or their dogs' behaviors? "That makes the most sense to me because I love Bernie and want to keep him," says Sonja. Because she can't afford a professional trainer, she embarks on a program of self-education. First she defines all of Bernie's problem behaviors, when they occur, and any possible reasons for them. Quickly discovering that she knows little about dogs and training, she reviews a wide selection of texts that discuss classical wild-dog behavior, traditional and innovative training techniques, and the relationship between owner and pet. Although she originally postponed any kind of remedial program, fearing Bernie had so many problems she could never train him properly, she soon realizes that all his misbehaviors relate directly to his stress when left alone. If she can relieve that stress, she can probably solve his (and her own) problems.

After considering many different stress-relieving techniques, a combination of denning plus confidence-building training and neutering appeal to Sonja most. Denning offers a quick fix: By confining Bernie to a fiberglass kennel with a special blanket and toy when she's away, she eliminates those hours of mulling over her dog's destructive behavior and those erratic emotional homecomings. She initiates daily obedience training and has Bernie neutered to help ensure that the denning need not be permanent. Much of the tension Bernie experiences while alone results from fear—the more confidence he gains, the less fearful he will be; the less fearful, the less tension and stress he needs to dissipate. For the dog who must spend a great deal of time alone, basic training builds essential self-confidence. Dogs who respond to commands, who work, who synchronize their brains and bodies, who succeed, take control of themselves and their emotions. They don't jump every time a door slams; they don't howl at every siren; they don't cower every time the phone rings. Consequently, when they're left alone, they sleep contentedly or play with a few favorite toys.

The choice to neuter Bernie took time for Sonja to accept because this was an unheard-of practice on her parents' farm. Sonja had to realize that she and Bernie don't live on the farm but in a city, where an isolated intact male—in a building with numerous other animals within hearing, smelling, and sometimes even seeing range, but with no way physically to contact those others—experiences cruel levels of tension and stress. Under these circumstances it's far more humane to neuter the animal and at least eliminate the sexual stress.

Within a month of implementing her changes, Sonja celebrates the noticeable improvement in her relationship with Bernie. Because she no longer deals with nightly messes, her responses to her pet become more consistent and assured. She downplays her exits and entrances to further accentuate their new low-keyed, mellow, and consistent interaction. Being a naturally intelligent and perceptive animal, Bernie responds beautifully. Formerly frantic greetings become equally joyful but calmer exchanges. By the end of the second month of the program, Bernie adores his training sessions and seems openly disdainful of rowdy behavior in other canines. Although he enjoys the company of other dogs, the neutered spaniel never indulges in territorial disputes, strolling casually past frantic males sparring over females in heat. Three months after she initiates her changes, Sonja can leave the door of Bernie's kennel open while she's gone. She toys with the idea of getting rid of the crate altogether, but Bernie seems to enjoy his little haven, sometimes even sleeping in it when Sonja's at home. By the time she and Bernie celebrate his first birthday, their first scared and lonely months together in New York seem like a faint bad dream.

But what about the Mathesons? How can they change Fezziwig's subtle and perhaps even imagined body language? The fact is, they can't. They can only change themselves; they can stop interpreting that body language as both significant and negative. Easier said than done, though, because lonely, bored, isolated humans so often and so readily zero in on all aspects of their dog's expressions and then assign them great meaning. If they make positive or at least nondetrimental connections—"Scoobie always curls his left front paw inward when it's going to rain," "Sandy licks me twice when it's time for dinner"—the owners hardly have a problem pet. But when their assumptions or projected emotions resulted in owner guilt, resentment, or other negative emotions, the relationship inevitably deteriorates.

Fortunately the Mathesons are naturally social people, and even though they have stopped going out, that hasn't kept them from inviting others in. As a progression of guests passes through their home, they come to realize one very important fact they'd previously overlooked: Both A.J. and Carla had lost sight of the fact that Fezziwig, at fifteen, had become an old dog. What they had interpreted as his expressions of loneliness and rebuke in response to their comings and goings were actually those the aged dog used to accommodate natural changes in his hearing, eyesight, and mobility.

The real revelation came when the entire family gathered for the holidays and Carla sadly commented to her daughter how Fezziwig gets so upset when they leave him that he refuses to jump on them in greeting when they return. Her daughter laughed, "Good Lord, Mom, old Fez hasn't jumped on *anyone* for years!"

Eventually the Mathesons exchange one set of beliefs for another, abandoning the idea that Fezziwig doesn't want them to leave him alone in favor of the simple truth: Their beloved Fezziwig isn't the same dog he used to be.

In many situations involving loneliness, we can eliminate the body language associated with these emotional states and our responses to the behavior by doing away with the environmental conditions that cause them. However, a surprising number of owners will listen to a dog howl, dig, or scratch in the kitchen or outdoors without ever considering that they could simply permit it to be with them. Such owners usually give two reasons for not allowing the dog in the house or bedroom:

- "I'm afraid the dog will destroy that area, too."
- "I don't want to give in to a dog."

Recognizing the social nature of dogs, we can easily see how confining them near, but without direct access to, other humans or animals produces a very potent form of stress and tension. The more social the animal, the more frantically it tries to get out of its enclosure and joins its "pack" or notify them of its whereabouts via barking, howling, or whining. Unfortunately many owners interpret these intense expressions as proof of the dog's *anti*-social and destructive nature. Thus owner and dog get caught in a vicious cycle of isolation leading to destruction, which leads to more isolation and more negative behavior, *ad infinitum.* If you routinely tie your dog in the yard or confine it in the kitchen, where it makes a racket or mess while you're trying to enjoy a quiet evening at home, let your pet join you. Sure, at first Carbuncle may be so ecstatic to be with you, he'll behave like a complete fool and seem capable of manifesting your worst fears at any second; but give him a chance to settle down. Gradually introduce some simple obedience training to increase his (and your) self-confidence and to let him know his place in your pack. You don't use consistent training just to establish your authority; you use it to enable your dog to recognize and be assured about its position in your household. Such awareness increases the dog's confidence even more and simultaneously decreases stress, tension, and related negative body-language expressions.

If we can't change either the environment or ourselves to get rid of the boring, frustrating, or isolating conditions or the stress they create, we must wrestle with the final choice: getting rid of the dog. Because body language related to these states includes such a panoply of negative behavior, owners who experience it are often driven to this extreme. Isolation-related misbehavior is probably the most common reason for taking a healthy pet to the pound or having it euthanized; yet few owners openly admit this fact. If owners work or spend a lot of time away from home, isolation behavior attacks their relationship with their pet with all the vengeance of a chronic illness. Unlike dogs who bite children or chase cars and give owners dramatic, generally acceptable reasons to call it quits, isolation behaviors seldom create such drama and rarely threaten anyone. However, they relentlessly grate on a relationship until they wear down even the most ardent dog lover. One week it's a favorite lamp shattered, another it's a warning from the dog officer about Jupiter's barking, the week after that, shredded newspapers—and on and on it goes. Only a truly insensitive cad would have his or her dog put down because "He chewed my slipper" or "She knocked over my plants." But when such minor events become one in a sequence that seems doomed to proceed as relentlessly into the future as it wound its way through the past, the result can be termination.

THE TERMINATION TRAP

Unfortunately most owners choose to terminate a relationship involving isolation-related misbehavior based on ignorance rather than knowledge. Initially they view the behavior as too minor or insignificant to seek professional help, or if it's a nondestructive but nuisance behavior, such as barking or whining that occurs in their absence, they may not even be aware of it until disaster strikes. After enduring often heartbreaking love-hate relationships for months or even years, one day they decide that they just can't take it anymore. Inundated with grief and guilt, they take the dog to the pound or the veterinarian to be euthanized. If such owners believe the dog's infractions to be relatively minor when considered individually, they often feel victimized by circumstances and society. Furthermore, even though they believe they're at the end of their rope, these people did at one time love and want their pet; to recognize that this is no longer true and that they're terminating the relationship for what others may consider poor reasons can cause extreme anxiety. Consequently, many owners in this position make up more convincing reasons to get rid of the dog: "The new baby's allergic to her." "I'm being transferred and can't take him with me." "When she knocked down my grandmother, my mother said she had to go."

While such excuses may seem extreme, more than one owner has found honesty to be less than the best policy when

explaining to those who aren't familiar with the trials and tribu-
lations of isolation misbehaviors just why they're making this
choice. For example, Tony Blanchard has never effectively
coped with his bearded collie's destructive isolation behavior in
their three years together. Tony views his dog's misbehaviors as
an attack by one species on another, and the negative effect on
his ego and his subsequent anger, violence, guilt, alienation, and
remorse steadily increase with each encounter. Because the dog
receives such regular and thorough punishment, she now
cowers and trembles whenever anyone comes near her. When
Tony discovers a large hole chewed in his leather jacket, he runs
out of the house and spends the night with a friend lest he kill
his pet.

"I'm going to have to put her down," Tony says sadly to his
friend. "She's scared to death of everyone. No one else would
take her, and I hate myself for yelling at her and beating her all
the time. What kind of a life is that?" His friend agrees, noting
that the dog's life can only get worse now that Tony's working a
lot more overtime.

The next day, a very subdued Tony takes his pet to the Sun-
nydale Animal Hospital. He expects to have to sign some sort of
release; he doesn't expect a third degree.

"Oh, what an adorable dog," exclaims the receptionist.
"Does she bite?"

"No," says Tony, wishing she'd take the dog so that he could
get out of there.

"Then why are you putting her to sleep?" wonders the
young woman.

"Because she chews and scratches everything." Tony knows
he has spoken too defensively and struggles to control his
mounting panic.

"What have you done to make her stop?"

Tony is now emmeshed in a no-win situation. In addition to
his own agony and guilt, he must deal with that heaped upon
him by a stranger. In desperation he signs the release, throws his
money on the counter, and runs. Other owners react with less
hostility; still others let themselves be talked into trying "one
more time."

When exasperated owners succumb to well-intentioned

pressure from dog-loving friends and animal-care specialists, they may be doing neither themselves nor their dogs any favors. If they don't take the time to evaluate their commitment to their pet honestly and embrace often time- and energy-consuming ways to deal with misbehaviors *right now*, about all they succeed in doing is generating more heartache and postponing termination until even worse conditions occur.

Back in the days when I was ignorant of the insidious nature of isolation behavior and the special training techniques it requires, my "treatment" for owners consisted of a cheerful pep talk and traditional (and almost invariably useless) training hints. One Irish-setter owner wanted me to euthanize his dog on several occasions, but I always inspired him to "try harder." I recognized the dog immediately when it was brought in dead by the dog officer.

"It's a stray that's been wandering out by the main highway for several weeks. Finally got hit by a truck."

A quick check with the owner's landlady confirmed that the owner had moved and left no forwarding address: "Glad to be rid of him, too. That damned dog of his just about ruined my place!" Because I denied the owner the option of humane euthanasia and refused to acknowledge how those, to me, seemingly minor misbehaviors had damaged his relationship with his pet, I acted as an unwitting accomplice in the dog's eventual demise under less-than-humane circumstances.

THE BONDED SOLUTION

Can busy owners ever find a way out of this labyrinth of emotions? The best solution is always prevention. Deal with each problem as it arises, don't let them accumulate. There's no such thing as an insignificant problem: If it bothers you, it's a problem. Maybe Rory only chewed an old sweater you meant to throw away; the point is, does it bother you? Does it make you feel that your dog is acting mean or spiteful or that you're somehow an inferior owner? If so, resolve to deal with the behavior now; don't allow it to erode your relationship anymore.

What if your dog displays behaviors that already test your

limits of endurance? We already know the first step: Get rid of the guilt. It's not your fault or your dog's; Rory is responding to an *environmental* state, not your treatment or mistreatment of him. Keeping that in mind carries us automatically to the second step: Dissociate all other negative emotions from the dog's behavior and your response. Rory isn't mean, spiteful, or stupid, nor are you. If you choose to attribute any of these qualities to either your dog or yourself, recognize this as a conscious choice you make and one only you can change. Now consider all four options, selecting the one that fulfills your needs and is therefore the one you can consistently implement without guilt or resentment for however long it takes to solve the problem.

As we'll see time and time again, bonded owners aren't perfect humans who own perfect dogs. What makes them special is their willingness objectively to evaluate the relationship with their dogs and make necessary changes to preserve its quality. The wide range of negative behaviors associated with boredom, frustration, and isolation present us with unlimited opportunities to extricate ourselves from detrimental definitions, emotions, and responses. How we respond to any problem behavior always involves making choices. We may choose to believe that Rory goes on a destructive rampage when left alone because he's mean, spiteful, and stupid. We may choose to believe that Rory wrecks the house because he loves and misses us unbearably, in which case we become cruel, heartless, mean humans who deserve to wallow in our guilt. Or we may choose to see Rory's body language and our subsequent emotional interpretation and body-language expressions as proof of our dog's and our own feelings and beliefs about being in an isolated and boring environment.

The third bonded choice recognizes that both dog and owner respond to these environmental states—*and that they may respond similarly or differently.* Initially neither Sonja nor Bernie possessed the self-confidence and experience to define time spent alone as unique rather than "bad." Fezziwig Matheson coped quite well with isolation, whereas his owners projected their own adjustment problems on to him. Other owners prefer to be alone, while their dogs crave constant companionship. None of these orientations is wrong—and the bonded owner, always striving to forge a bond from the most acceptable orientation, recognizes this above all.

A pop maxim urges us to make lemonade whenever the world hands us lemons. Boredom-, frustration-, and isolation-related behaviors provide us with some of the bitterest encounters with our dogs. By removing the bite, the acid of negative emotion, and replacing it with self-confidence and knowledge, we can sweeten even the most frustrating relationship.

Problem isolation behaviors and our responses to them depend to a large degree on time. A dog left alone for a few minutes probably won't get as bored or do as much damage as one left alone all day. When our dogs do destroy property in our absence, we usually feel guilty about not spending enough time with them. Many other interactions with our pets are also directly or indirectly related to our sense of time: Is it time to begin housebreaking Monique? Is it time to take Herbie for a walk? Why does it take Peaches so long to find the right spot to relieve herself? Why can't Chico slow down?

In the next chapter we're going to explore how the human and canine senses of time can try the patience of both species.

9
PATIENCE AND IMPATIENCE: FINDING THE RIGHT TIME

IT'S TOUGH TO BE a Saint Bernard in the city. By the time Splinter celebrates his first birthday in a Denver animal shelter, a couple of hundred people who live within a fifty-mile radius of the suburb where he and his nine littermates were born would attest to that fact. "They're too big." "They're too active." "They're too lethargic." "They're too smart." "They're too stupid." Despite all the advice to the contrary, Jim and Jenny Nesbitt decide to give Splinter a home. "He sat so patiently while we looked at all the other dogs, I couldn't resist him," admits Jenny. The Nesbitts cram the somewhat hesitant but nonetheless cheerful dog into the backseat of their sporty sedan and head home.

"He's such a smart dog," notes Jim, impatient to get started teaching his new dog old tricks, "I should have him completely trained in a month or less."

On the other side of town Splinter's littermate, Effie, spends her first birthday alone in the Martinellis' garage. "Oh c'mon, Dad, can't we let her inside for her birthday?" beg the Martinelli children.

Lou Martinelli shakes his head. "No. That dog tries my patience. She's an outdoor dog, and she's going to stay outdoors. You kids know how crazy she gets when she's around food."

Although Lou likes Effie and would like her to be a house-pet, her hyperexcitability makes her impossible to handle. Whether the Martinellis are trying to feed, walk, or play with her, she always demands more, faster. Lou and his wife, Winnie, worry that Effie, given her great size, might accidentally hurt one of the children; and so they forbid them to go near the dog except in their presence.

From the kitchen window Winnie can see Effie lying on her rug in the open garage doorway. People and bicycles constantly pass their house, and countless scents and sounds permeate the air of their crowded neighborhood. So much to see and do, yet Effie seems quite content to lie and watch the world go by.

"Such patience," murmurs Winnie to herself. "Why doesn't she behave like that when she's with us?"

Even though patience and impatience aren't emotions in the strict sense of the word, they can precipitate a raft of positive and negative behaviors and emotions in both people and dogs. We join the Nesbitts in applauding Splinter's willingness to allow toddlers to climb all over him and tug his ears: "What a saint!" Then we cringe uneasily as he patiently sits in a downpour waiting for his owners: "What a fool—doesn't he know enough to come in out of the rain?" Owners who lack the patience to consistently train a young pup invariably find themselves wallowing in anger, guilt, and frustration every time their adult dog misbehaves. By the same token, impatient dogs can be a constant source of disappointment to their owners. Have you, like the Martinellis, ever commanded your pet to sit quietly while you place its food dish on the floor, only to watch your dog leap wildly into the air and send dish, kibble, and even kitchen chairs flying? And surely every owner has admonished the just-bathed pet, "Now stay right there just one second until I get a towel," only to see the soaking-wet animal roar into the living room to shake itself dry.

Patience and impatience don't involve an either/or state. We seldom view so-called normal dogs as patient or impatient, but rather apply these definitions to an exceptional relationship between the dog and something or someone in its environment. If Effie simply accepts and eats what's placed before her with no extraordinary display, we accept that as normal. However, when she lunges frantically at her food, we call her impatient. If the Martinellis forget to feed her and she does nothing to draw attention to her empty dish, we admire her patience.

Owners define patience and impatience in themselves in a similar manner. Jim Nesbitt can almost taste his desire to start working with Splinter the instant they get home. "Don't be so impatient," Jenny reminds him. When he senses frustration and anger welling up after Splinter ignores a command three times, Jim takes five deep breaths and tries to control his irritation. When his anger and frustration fade, he congratulates himself on his patience. If Jim gives a command and sees Splinter respond immediately, he simply accepts the exchange as normal.

While these examples might seem to suggest that most definitions of patience and impatience are capricious and variable to the point of meaninglessness, in fact both states clearly relate to time—or, more precisely, to the owner's sense of time as it relates to the dog's.

WHAT'S TIME TO A DOG?

Although the universe began millions of years ago, most people operate under the assumption that "Time began when man began." Time is such an intimate part of our lives that we use it to measure and define our most basic activities. Yet, as crucial as time may be to our day-to-day existence, we often fail to see that it is *our* time and in some situations may not coincide at all with the canine clock.

Whether a second or minute means the same to both dog and owner matters less than whether they find a certain time interval mutually acceptable. As long as Splinter relieves himself as soon as the Nesbitts let him out at night, the actual length of time it takes doesn't matter; and if Effie sits quietly as long as the Martinellis want, her behavior passes muster.

In other words, as long as a dog's behavior, or a human's response to it, doesn't strike us as being too fast or too slow, we don't care about the compatibility or incompatibility of canine and human time frames. Only when the two species diverge in their definitions of what is acceptable do problems arise. How do you feel when you're waiting for someone very special and he or she stands you up? I usually became sad and anxious, then depressed, frustrated, and angry. Hidden-camera studies of dogs primed for enthusiastic homecomings indicate that our canine companions experience similar emotional tensions when expected events don't occur at expected times. In these experiments the dogs were first trained to a classic sequence of exuberant response to their owners' returns. The dog barks and dances toward the owner, who encourages it with sweeping gestures and joyful cries. The dog leaps into the air; the owner catches it and/or hugs it. The entire display usually takes less than a minute, but it contains a wide range of body-language signals and emotions.

Having so conditioned the dog, the researchers delay the expected homecoming and observe the results. As the moment of the anticipated rendezvous approaches, the dog's tension mounts. It becomes restless and moves toward the door or peers out the window. When the owner doesn't appear, the dog's anticipation gives way to anxiety; more dependent animals become fearful. All but the most stable and well-trained resort to some form of isolation behavior—chewing, whining, barking.

Obviously our dogs don't depend on the kitchen clock to tell time; yet their time awareness is every bit as acute as mine as I stand beneath the huge clock on the town hall waiting for my tardy friend. As soon as we link a specific time or interval with a specific event, we attach a very special meaning to time. Having made our dogs and ourselves aware of the special meaning it then becomes an inextricable part of our relationship.

The experiment also demonstrates the role consistency plays in establishing human-time awareness in our pets. Had the researchers not conditioned the dogs to the same event at the same time, they would have produced entirely different results. Random demonstrative homecomings initially result in a dog who is constantly on edge, a dog that has no apparent sense of time with response to the homecoming event. Because the dog can't link time to the event, it can never relax. If it can't adapt to the constant vacillation between the extreme negative tension of isolation and the equally extreme "positive" tension of the homecoming, it copes by responding to both in the same way, either by remaining in a constant state of agitation or by ignoring both once highly charged events.

Therefore, consistency not only facilitates learning by establishing a predictable event on which the dog can count, it simultaneously establishes a time rhythm or harmony between owner and pet. The owner gives the body-language and emotional cues to which the dog responds within and/or at a specific time and *vice versa;* the entire sequence appears as a unit to dog and owner alike. By the same token, inconsistent timing, such as when Jim expects Splinter to respond instantly today but allows him longer tomorrow, gives the dog two "acceptable" times in which to fit the same sequence: Which one is right?

Imagine traveling to another planet where alien beings measure time in argiles and expect you to perform gleams, but your human vocabulary and experience include neither the word *gleam* nor the concept of an *argile.* However, when the alien says, "Gleam!" it waves a tentacle whenever you move toward it, and your movement makes the creature quite happy. In such a way you learn to recognize a gleam as movement toward the alien. Depending on how fast or slowly you move, the alien responds more or less positively. In such a way you discover both the proper response and the desired time or rate in which to accomplish that response: You begin to intuit the meaning of an argile.

Think about that. During the process you learn to do more than just walk toward the other when it utters, "Gleam." You also learn that this means that you approach at a specific rate—a brisk trot, for example; and while the alien may judge your rate too fast, too slow, or just right in terms of argiles, you still respond in terms of earthly seconds. In addition, you begin to assign emotions to the different expressions the alien uses to indicate you're approaching too fast or too slowly.

Once you master both the desired activity and the preferred rate of performance, the alien asks you to perform your display less often. In fact, once you can do it flawlessly, the occasional request actually sharpens your presentation as you try to elicit the creature's happy responses.

MY TIME IS YOUR TIME

The same phenomenon occurs with our dogs. By teaching them a particular sequence of behaviors in a particular time frame, we create a multilevel form of communication that includes emotions as well as body language. To elicit a consistent response from our pets, though, we must initially be consistent in our demands. If Jim wants to teach Splinter to come, he needs to clarify his own idea of what that means *before* he can convey it to his dog. Only then can he reinforce the canine display that most closely approximates his idea. If Jim has a vague or variable idea of what he wants from his dog or the time in which he expects the proper response, any reinforcement or punishment is meaningless. Not only does the correct response elude the dog, he never learns what's too fast or too slow. Eventually Jim will feel out of synch with his dog, growing increasingly impatient with both his pet and himself as a trainer.

Similarly, the dog whose owner reinforces or punishes a particular sequence inconsistently also becomes frustrated, bored, or worse. One Saturday, the Nesbitts take Splinter for a romp in the woods. In response to their call the dog comes flying, knocking them down in a joyful greeting that Jim and Jenny find hilarious. The next day the Nesbitts put Splinter out to relieve himself before they depart for an elegant dinner party. "Splinter, come," calls Jim, as he adjusts his tie. He's shocked and furious when the Saint Bernard bowls him over. Two days later Jenny calls Splinter in from the backyard just as the phone rings. During the ten minutes Jenny's engrossed in her call, Splinter wends his way to the door. "Good dog!" Jenny gushes as she lets him in.

While the Nesbitts may believe they're teaching Splinter to come when called, they've done little if anything to achieve that goal. While Splinter may have attached a specific meaning to the word *come*—that is, move toward the person who utters that word—his owners' inconsistent response makes it impossible to know when or how fast to exhibit that behavior. When he dashes to them, they reward the behavior one day but punish it the next, thereby destroying Splinter's ability to make any consistent time event connection. Jenny praises the dog for wandering toward the door in ten minutes, but how likely is it that Jim will respond similarly when he calls the dog in from the freezing rain? If the alien says, "Gleam!" and delights in your approach one day and punishes it the next, you must question whether its *Gleam!* and your approach behavior are related at all. If it smiles broadly whether you approach at a brisk trot or a snail's pace, you're comfortable with the meaning of the command, but attach no time interval to the proper response. While consistent behavior reflects patience, inconsistent behavior arises from and leads to canine and human impatience. The only way we can hope to make our time our dog's time is to reinforce our concept of time consistently.

Given our desire to create a bonded relationship, shouldn't we also strive to accept canine time as our own? We should, but only the most confident and consistent owners can succeed because it requires we be intimately aware of our pets. For example, some dogs, like some people, naturally move slower or faster than others. These differences may be related to breed, physiology, the specific event, or a host of other variables. As a group, miniature and toy poodles are more physically demonstrative and excitable than the more phlegmatic bloodhounds, although bonds with their owners may be equally strong. An older dog with arthritis may be physically unable to respond as quickly to a come or sit command as a younger dog with sound hips. A dog who hates being brushed but loves going for a walk responds slower or faster to its owner's commands, depending on its desire to participate in these activities. If the owner is unaware of these variables, he or she may choose to believe that the dog is deliberately snubbing the owner's sense of time.

WHAT DIFFERENCE DOES IT MAKE?

It doesn't matter whether we accept dog time all the time ("Do whatever you want to do, it's fine with me, Buster."), expect our dogs to respond only according to our time ("When I say jump, you jump. And not one second sooner or later!"), or

recognize some combination of the two. As long as our definition works, and as long as we can consistently reinforce the individual sequences within our chosen orientation, the relationship
will be stable.

In my experience, however, both the all-dog and the all-
human approach to timing cause more problems than the blend
simply because one-sided views tend to be so rigid. For example, Chloe Haversham dotes on her Lhasa apso, Bendix, a dominate male who more or less runs the Haversham household.
Initially Chloe accepts everything Bendix does. Any demands
she makes, she presents in ways that reflect what the dog was
going to do anyway. As he drags her out the door, she "commands," "Let's go for a walk, Bendix." While we earlier proposed setup training that exploits the dog's natural tendencies
as an excellent method to build confidence in the shy dog, Bendix already has more confidence than he knows what to do with.
In fact, he's downright arrogant and rude. In this situation,
Chloe's willingness to let Bendix respond when (and if) he
wishes basically communicates submission. She's not being a
sensitive, patient owner responding to his special needs; she's
simply abrogating her responsibility to accept dominion over
her bossy dog.

Furthermore, having acceded to dog-time, Chloe must adhere to it herself. By patiently allowing Bendix to do things in
his own time, she reinforces his bad behavior. Therefore, when
the Lhasa demands or expects something of her, it should come
as no surprise that he wants it in *his* own time. If he's hungry, he
wants to eat *now;* if he wants to go out, he wants to go out *now.*
If he doesn't want to be groomed, he wants her to stop *now.*
And so in the process of responding patiently to her impatient
dog, Chloe finds herself involved in a relationship whose pace
eventually makes her resentful and, yes, impatient.

Lou Martinelli reaches a different but equally unnerving
impasse with Effie. He believes a well-trained dog should always respond exactly as the owner has taught it. While such a
simplistic approach would seemingly guarantee consistency, it
doesn't allow for a dog whose time sense differs from its owner's
in certain situations. For example, the Martinellis previously

owned an Irish setter that Lou gave away because its hair-triggered response to everything drove him crazy. The dog was simply too fast (impatient), and trying to get it to slow down and accept human time degenerated into a battle of wills. In general, Effie's timing more closely parallels Lou's, or at least it did when she was a young pup. When she was four months old, however, Lou started working overtime, forcing a halt to their daily training sessions. At the same time the children went back to school, and Winnie became involved in outside activities. Consequently the interactions Effie had grown to expect became infrequent events, and because she once enjoyed them so much, she suffered great emotional consequences. The greater the positive emotional charge an animal gets from an interaction, the more anxiously it wants to experience that event again. Thus, when Lou attempts to resume training the eight-month-old dog where he left off at four months, he finds her much more excitable, and seemingly oblivious to his commands. This he erroneously perceives as evidence that she doesn't want to learn and doesn't want to please him.

The Nesbitts' orientation creates an entirely different kind of problem for them and Splinter. Jim pursues such a vigorous "Jump when I say jump!" campaign that he soon overwhelms his dog. At first the training totally confuses Splinter, but Jim's consistently firm punishment quickly spurs the Saint Bernard to make some rudimentary connections to avoid being chastised. Rather than risk punishment by doing something "wrong," Splinter decides to do nothing at all. Even when engaged in an activity as mundane as eating or chewing on his toys, he constantly looks to his owners to see if his behavior meets with their approval: Is he eating too fast? Too slow? Are they going to ask him to come, sit, or stay? His owners' rigid sense of time forces him into a perpetual state of uneasiness and, quite frequently, immobility.

By now we can appreciate how time incompatibilities affect the quality of the relationship. Owners who initially view themselves as patient come to see themselves as abused by their dogs; this leads to anger, resentment, and even impatience. Others who believe their dogs respond too fast or slowly also become impatient with their pets. They, too, feel anger and resentment; and if they react negatively and excessively, they also suffer guilt.

Because such problems affect the most fundamental relationship between owner and pet, simply getting the dog to do what we want, in the time and manner we want, may not noticeably improve the situation. Although the Nesbitts can say that Splinter is becoming trained, both of them find the Saint Bernard's attitude disconcerting: "I wish he'd act like he enjoys being with us more," sighs Jenny. "We love him so much, I can't understand why he seems so uneasy all the time." Meanwhile the Martinellis increasingly realize that training Effie to sit and stay isn't going to gain them much satisfaction if she sits poised like a cannon about to explode. What these owners want isn't just a dog that behaves; like all owners, they want a dog that *wants* to behave.

BATTLING BENDIX

While we usually think of patience as "good" and impatience as "bad," patience can actually be detrimental in certain situations. Let's take a closer look at the relationship between Chloe Haversham and Bendix. Chloe's patience is legendary in the community. While the rest of us slink into our closets when requested to drive nine rambunctious Little Leaguers to a game or sew 300,000 sequins on a centennial banner for the volunteer fire department, Chloe always leaps at such opportunities. Therefore, when Bendix joined the Haversham household one Mother's Day, everyone thought Chloe's patient nature would quickly triumph over the feisty furball. Instead of improving the Lhasa's temperament, however, her undemanding disposition only made it worse.

Although she eventually found herself in a submissive-owner/dominant-dog relationship, that's not at all that Chloe had intended. At first she believed she could calm the dog and win his obedience with her patience. After all, hadn't she been able to settle down her boisterous son and his friends by patiently ignoring their demanding behavior?

To be sure, such an approach can work with animals whose

concepts of time more or less match their owners'. We noted in the last chapter how Sonja extinguished Bernie's jumping simply by not acknowledging—punishing *or* rewarding—it in any way. But Bendix marches to the tick of a different clock. Furthermore, while Chloe patiently ignores Bendix or, worse, tries to avoid upsetting him, he remains quite impatient about having his own needs fulfilled; and it doesn't bother him in the least to pester her.

In such a way owner and dog create a highly unstable, incompatible relationship in which neither gets what he or she wants. Chloe wants a well-behaved, loving pet; Bendix is arrogant and aloof. Bendix wants an owner who will do his bidding; sometimes Chloe does, but more often she ignores him. As time goes on, Chloe becomes resentful that all her patience hasn't produced the loving pet she wants, and Bendix grows more persisent and dominant in his displays in his efforts to gain her attention.

Essentially Chloe and Bendix want the same thing: the other's love and attention. However, because one seeks to gain this with patience while the other uses impatience, the probability of achieving either goal approaches zero. Moreover, as long as Chloe responds to Bendix's impatience with even more patience, she reinforces the gap between their definitions of time. Unfortunately, because we tend to consider patience a more praiseworthy, elevated, and even saintly emotion compared with impatience, it's often hard for owners like Chloe to realize that their orientation might contribute to the problem.

"You mean I should clobber Bendix when he expects me to let him out the instant he barks at the door?" No, but obviously Chloe's present approach isn't working, either. Because Bendix is so dominant, Chloe's boundless tolerance has simply created a dog who barks longer and louder. In two months Chloe has extended her patient response far beyond her initial five minutes, now calmy listening to him bark for over an hour before she gives in and lets him out.

Unfortunately set-ups that allow Bendix to succeed backfire because he already believes he can and should have whatever he wants whenever he wants it. Chloe might decide that rushing to open the door to let the Lhasa out *before* he barks would denote control over the dog. However, because Bendix wants her to open the door as fast as possible, her set-up only reinforces his problematic belief. Not only does Chloe fail to achieve the control over Bendix she wants, she winds up responding much faster to him than she finds comfortable. Whereas formerly she tried to slow his time down to match hers by ignoring him, now she tries to speed up her own to complement his. If neither dog nor owner sees any good reason to adapt to the other's time, any training technique will fall short of the desired results.

When mismatched senses of time are problematic, the best solutions arise from establishing a compatible and mutual recognition of the other's time. Suppose that when Bendix goes to the door and barks, Chloe remains seated and commands the dog to come to her instead. When he responds to this command, she praises him. Shortly, but not immediately, thereafter, she goes to the door and offers him the choice: "Want to go out?" In this sequence the impatient dominant dog receives the following messages:

"No, I won't jump when you say jump." *Then*
"I'm the one who's in charge here." *And finally*
"I know you have needs too, and I want to recognize them when I can."

By giving Bendix an alternate command in response to his demands, Chloe gives him the attention he wants but in a different way. Because the attention carries no negative connotation, it's easier for him to respond to her positively. (Compare this with the owner who screams at the barking dog to get away from the door, or smacks it and drags it away.) When Bendix responds appropriately, she praises him, giving him more attention but again on her own terms and in her own time. By then providing the dog with the opportunity to do what he wants, but within a more compatible time frame, Chloe communicates her willingness to fulfill his needs by choice. In other words, he doesn't have to force her to let him out by being obnoxious. Consistently substituting "my time" sequences followed by recognition of "dog-time," Chloe can temper her dog's impatience in a way that's consistent with her normally tolerant nature.

PATIENCE TO A FAULT

Another problem associated with patience occurs when mutually tolerant owner and dog essentially convey the same message to each other: "I don't care what you do as long as you don't bother me." Unlike the patient and impatient combinations, which may create positive and negative human/canine tugs in all directions, this combination creates virtual paralysis. The owner accepts the dog as simply there; the dog acts as though it feels the same way about the owner.

Such a breakdown occurs between Winnie Martinelli and Effie when Lou banishes the dog to the garage. From that time on, Winnie experiences virtually no close contact with Effie. Lou feeds her, takes her to the vet's or out for an occasional walk; most of Winnie's interaction consists of talking to or observing Effie from the kitchen window. Sometimes Effie wags her tail in response to Winnie's words; sometimes she acts as if she doesn't even see or hear Winnie at all. Other times Effie barks and wags her tail, but Winnie stares out the window without acknowledging the Saint Bernard's presence. In these cases, Effie may wag her tail once or twice, but quickly gives up when Winnie fails to respond.

Although we can't classify such a relationship as bad, we can't call it good, either. Because no bond exists between the parties, it's really a nonrelationship. The dog could run away or be given away and, aside from a few brief regrets during the actual parting, the owner would suffer no particular loss. Similarly the dog would leave the household or accept a new owner without a backward glance.

While those who dearly love their pets may find the idea of such a nonrelationship repulsive, such alliances aren't that uncommon and generally occur when people acquire a dog for some reason other than a deep desire to relate to it. Those who get dogs "for the kids," "to keep Mother company," or "for protection" seldom feel sufficient commitment to bother recognizing any but the animal's most basic needs. While they don't abuse their pets, their dogs are little more than animated pieces of fur-covered furniture, passively accepted as long as they don't complicate their owners' lives.

Because such owners usually see their interactions or noninteractions with their pets as perfectly normal, they seldom improve their relationships. If nothing's "wrong," there's nothing to fix. Usually it takes some dramatic event to point out that things could be better or that the relationship means more than the owner has previously acknowledged.

For example, one day Winnie must rush to the post office before it closes. Racing to the garage, she jumps into the car and backs out of the driveway, thinking only of her errand. Effie's yowl of pain cuts through her like a knife. Just inches from the rear wheels, the Saint Bernard sits quivering, one paw held limply, her great brow furrowed and her big brown eyes fixed on Winnie. Winnie leaps from the car and dashes to the dog, sobbing and trembling so badly she can hardly examine the injured foot. As if to reassure her owner, Effie licks her hand. In that instant "just a dog" becomes a very special one.

In this sort of situation, the owner's patience doesn't particularly benefit the dog. Both Chloe and Winnie use their patient natures to distance themselves from their pets, but don't see their orientations as detrimental. While Chloe uses hers as an excuse, a way to avoid accepting dominion over a canine tyrant, Winnie uses hers to avoid deeper emotional involvement with her pet. To be sure, Winnie might think her patience enables her to accept canine behaviors others might find infuriating and punishable, but she also resists initiating or accepting any *positive* interactions with Effie, too.

During the time it takes for Winnie to rush Effie to the vet's, wait for the X rays and diagnosis, and return home, her entire attitude toward her dog undergoes dramatic change. She unhesitatingly spent over a hundred dollars the Martinelli budget could ill afford on a dog she thought she didn't care about one way or another. Suddenly she realizes she does care: "Lou, we've got to do something about Effie. She shouldn't have to spend her entire life chained outdoors."

THE IMPATIENT OWNER:
CONSISTENTLY INCONSISTENT

Because impatient owners are by nature inconsistent, their dogs rarely respond in a consistently patient or impatient manner. One day the owner becomes furious because the dog moves too slowly, the next because it reacts too fast. Because of this, owners who find themselves losing patience with their pets must take the time to form a clear idea of what they really want to accomplish in the relationship.

When Winnie decides they should do all they can to include Effie as an intimate part of their household, both she and Lou define Effie's mealtime behavior as the major obstacle to overcome. The dog seems so anxious for her food that she begins whining and carrying on an hour before feeding time. Whoever carries her bowl to her must approach carefully so as not to be flattened by the big dog in her excitement; lately the younger children act quite cautiously around their pet. Whenever Effie acts this way, Lou feels a surge of anger; sometimes he yells at her, other times he smacks her, and on several occasions he refused to feed her at all to "teach her a lesson."

Initially Lou decides to solve the problem by teaching Effie to respond to a sit-stay command regardless of who gives it. That way family members could make her sit while they placed her dish some distance away, then move off before she pounced on her food. However, on the off chance that Effie might have some hidden medical problem, Lou calls his veterinarian to discuss the Saint Bernard's behavior and his planned solution. While the vet agrees that Lou's plan would work, he suggests an alternative: feeding Effie several smaller meals throughout the day. "This would have several benefits," the vet claims. "First, she wouldn't experience so much tension as feeding time approaches. Second, she may be able to utilize her food more efficiently and decrease those hunger pangs. Third, she'd have more interaction with her family. And finally, such a plan decreases the probability of her developing some serious digestive problems." The vet also recommends eventually putting Effie on free-choice feeding (having dry food available at all times so that she can snack rather than eat one or two big meals) to deemphasize the feeding process even more.

The Martinellis immediately implement the proposed changes, and within a few months Effie's impatient mealtime frenzy is transformed into contented munching. This softened behavior makes her so much more appealing to the family that by her second birthday it's hard for them to remember she had once been strictly an outside dog.

What Lou originally perceived as impatient behavior, which he consequently reacted to impatiently, really resulted from a combination of tension-producing physiological and behavioral conditions. Introducing multiple feedings not only relieved the hunger-related tension, it also increased the family's daily interaction with their pet and decreased her isolation. By removing these two major sources of tension, the family eliminated the impatient feeding-time display.

However, suppose Lou had decided that training or altering Effie's feeding schedule amounted to giving in to the dog. Suppose that he believed her failure automatically to sit patiently at mealtime indicated her spiteful and defiant nature. If so, both he and Effie would be happier if she went to a different home. In this scenario Lou would be responding to what he considers Effie's stubborn impatience with an equal amount of stubborn intolerance. He can't accept her behavior, yet he considers it demeaning to alter it or his own. Consequently, the relationship would be at an impasse, and both owner and dog could only hope to wear away at each other until one or the other eventually changed. While beneficial change can occur in this rather roundabout manner, more often the owner's anger and resentment increase as the dog's behavior persists or grows worse. And over time the negative effects could become permanent.

Across town, the Nesbitts' impatience leads to another chain of familiar problems. When they first adopted the year-old Saint Bernard from the pound, most of their friends shook their heads in disbelief. "Do you realize how much time and patience it takes to train an older dog?" Jenny's mother demanded. "You don't even have the patience to keep the oven door closed until a soufflé's baked!" Jim doesn't fare much better: Anyone who's ever worked with him or ridden with him in a car will attest to his impatience.

While most of us cherish patience in our dogs and consider it a virtue in humans, we often accept our own impatience as a way of life rather than a curse. Jim and Jenny view their impatience as an acceptable outgrowth of their youth, energy, intelligence, and enthusiasm. There's so much they can and want to do, they simply haven't either the time or desire to sit around waiting for soufflés, traffic lights, other drivers, or dogs. Their attitude isn't so much belligerent or arrogant as it is a reflection of their beliefs that "We're going places, so don't get in our way!"

Enter the lovable, galumphing, low-keyed, and less-than-perfectly trained Splinter. The resulting clash of Nesbitt Standard Time with Splinter Standard Time produces an emotional jet lag that leaves the Nesbitts reeling. If a particular training method doesn't yield positive results in a week, they shift to another. One day they discipline with a loud "No," the next day with a swat, and on the third they use distraction. Fifty pounds of commercial multicolored semimoist dog-food nuggets shaped like Saint Francis give way to Mother Nature's Organo-Burgers, then Dr. Doolittle's Swiss Army Dog Food. Although Splinter's problems aren't unbearable, the impatient Nesbitts are no closer to solving them six months later. In fact, they've even added a few more.

"I wish you'd be more patient with him," pleads Jenny one evening after Jim has exploded at the dog for the second time in as many days.

"It's not me, it's that stubborn mutt. Maybe I'm not cut out to own a dog."

Many problems arising from or resulting in impatience spring from different attitudes about time, not from any genuine canine or human deficiencies. If the Nesbitts could learn how dog-time differs from human-time and that dogs require consistency, they would realize that their multiple training methods and diets can't possibly accomplish lasting benefits. They simply aren't giving Splinter enough time to grasp the meaning of a command or disciplinary technique, let alone learn to respond to it. Similarly, a week or even two on a diet barely gives the canine digestive tract time to adjust to the change, let alone respond positively or negatively. In Splinter's case, the continual shift from one food to another actually creates more problems.

If Splinter wants to do something or be somewhere sooner than his owners, the Nesbitts would also profit from analyzing the behavior in terms of time differences. Imagine trying to read something written in an unfamiliar foreign language, then switching to something written in your native tongue. You read the former much more slowly than the latter and may need to read the same phrases several times before you're sure you grasp their meaning. When the Nesbitts attempt to teach Splinter to respond to commands or recognize the meaning of discipline, they're essentially asking him to understand a foreign language, and that initially requires more time. Conversely when Splinter races to the park, they should realize he's responding to a very familiar scent language that is quite foreign and even incomprehensible to them. Consequently, he responds much more rapidly than they, if they're even willing to respond at all.

Unfortunately, and like many owners, the Nesbitts believe their comprehension of the proper response to "Sit" or "Stay" is readily available to their pet. They also expect Splinter automatically to ignore all stimuli beyond human comprehension or respond in what they consider an acceptable (i.e., human) manner and time frame. Considering the major differences in how and what data is processed by human and canine sensory systems and the diverse ranges of body-language responses available to each species, such owner beliefs are bound to create problems. Unless we're aware of these differences, we may erroneously attribute our pet's failure to respond in the manner and time we consider normal to our own impatience (and all the negative emotions we link to that impatience) and/or to our pet's stupidity or spitefulness. Similarly, when our pets respond excessively or too rapidly to what we consider minimal or nonexistent stimuli, we could erroneously condemn them for their foolish impatience or berate ourselves for our ignorance.

SOLVING PATIENCE-RELATED PROBLEMS

Once we realize that a time problem exists, we can apply our four options. Accepting problems related to our impatience with our pets comes fairly easily if we analyze the problem(s) in terms of time differences. Rather than believing that they themselves are too impatient or that Splinter's too slow or fast, the Nesbitts can accept their own behavior and their dog's as different instead of right or wrong. Let's assume Jim and Jenny accept that their accelerated time sense prevents them from developing the patience they need to train Splinter properly. If they still want a well-trained dog, they can easily justify the expense of having Splinter professionally trained. In that case the owners' acceptance of the time difference helps them improve the relationship. When owners define problems in terms of the need for basic training, it doesn't matter who trains the dog as long as it get done.

If we think a given problem increases our own or our dog's impatience, changing our own or our dog's behavior can pose quite a challenge. For example, even though Splinter responds beautifully to his professional training, Jim and Jenny remain too impatient to provide the daily reinforcement necessary to establish permanently the dog's new patterns of behavior. Consequently Splinter backslides during the months following his training and is soon up to his old tricks again.

Although the Nesbitts could send Splinter back to the trainer, they decide that their own impatience needs as much attention as their dog's misbehavior. "If we're not going to do what the trainer says, we're just throwing money away," Jim concedes. Nevertheless, changing their sense of time to coincide more closely with their dog's needs necessitates major changes in their pesonal philosophies and life-style. This would be such a significant undertaking that the Nesbitts feel they must also weigh the option of termination.

However, Splinter is so lovable that neither Jim nor Jenny can even begin to consider termination objectively. "I feel simply horrible, like an absolute rat, every time he looks at me," groans Jenny guiltily. On the other hand, Splinter's behavior is worsening to the point that it threatens to damage their relationship irreparably. To reduce their anxiety and increase their objectivity, Jim and Jenny decide to board Splinter for a week at a local kennel and use that time to thoroughly evaluate the termination option.

When the Nesbitts sample life without Splinter, several previously unacknowledged facts come to light. Although they celebrate their first evening of freedom by going out to dinner after work because they don't have to rush home to let the dog out, they find the restaurant scene less appealing than before. "I've gotten so used to taking a long walk with Splinter before dinner, then relaxing with Jenny and him in the evenings," comments Jim as he waits impatiently for the waiter to bring the check.

The next evening, the Nesbitts stay home, but the house seems too big and empty without the lovable Saint Bernard. Jim makes a few feeble comments about how nice it is not to be dragged to the park twice a day, and Jenny halfheartedly agrees. By the third day, Jim and Jenny realize they no longer are, nor do they want to be, the fast-track couple of the future. Enjoying that certain element of homeyness and stability Splinter's presence brings to their lives, they realize they're willing to change in order to preserve a quality relationship with their pet.

The Nesbitts spend the remaining days of their trial separation evaluating at-home training techniques. Now that they recognize that they want to spend time with Splinter and that he's not keeping them from other, more desirable activities, they become excited about the prospect of consistently and patiently training their dog themselves.

THE BONDED APPROACH:
SYNCHRONIZING OUR WATCHES

What a highly charged reunion when Jim and Jenny pick Splinter up at the kennel! In their week without him they've realized how much more their relationship means to them than they ever imagined, and they're eager to alter their former impatience. But what about Splinter: Is he willing to alter his sense of time and meet his owners halfway? "We'll cross that bridge when and if we come to it," Jim declares, citing Splinter's joyful greeting as proof that the dog wants to be with them as much as they want to be with him. As the Nesbitts concentrate on consistently solving various problems during the following weeks, they forget about the negative emotions they believed caused the unacceptable behaviors. When Splinter learns to walk calmly on a leash and heel, Jim and Jenny abandon memories of the dreadful impatience they originally attributed to his unruly lunges to the park. They settle into a comfortable, consistent training routine with Splinter, and as they see him readily respond, their own impatience vanishes, too. What's occurring between Jim, Jenny, and Splinter is the formation of a lasting bond as the original tug-of-war between patience and impatience gives way to an awareness that each individual has his, her, or its own time.

Dealing with patience and impatience takes tremendous effort because the feelings surrounding these states are so ephemeral and all-pervasive. Losing one's patience is something like riding a roller coaster, plunging from serenity one moment into anger, despair, frustration, and guilt the next. That's why bonded owners invest all the effort it takes to create and maintain time-tolerance in themselves and their pets.

Whenever I find myself in a situation where patience or impatience with my dogs seems a crucial factor, I pose a question suggested by a former classmate: "In ten years, what difference will it make?" While some might argue that ten years is too long when talking about a dog with an average life expectancy of thirteen years, this rhetorical question does permit us to cut a lot of emotional red tape. If you believe that your dog's or your own response will be just as irritating and negatively meaningful ten years from now, you'd better do something about the problem now because it can only get worse. But if the irritating quality of the behavior depends mostly on a unique time and place, you might be doing yourself, your dog, and your relationship a big favor if you try to reduce the emotional charge and have patience.

Regardless of how we look at it, patience is a function of confidence. The more we believe and have confidence in what we're doing, the more relaxed we are about doing it. The more confident we are, the less threatened we are when a timetable doesn't work out exactly the way we wanted. Recognizing the role patience and impatience play in our relationships with our pets leads to an understanding of the tempering process that occurs during the formation of a strong bond. Few beings on this earth have the power to make me feel as impatient as my dogs; yet simultaneously, few have given me so many opportunities to learn patience, confidence, and self-control. Not only can we learn to respect our dogs' special clocks, we can teach them to respect ours. Even so, the differences, not the similarities, put the richness into most relationships. What a boring world it would be if everyone, people and dogs included, all ticked to the beat of the same clock.

In the next chapter we're going to take a look at the brighter and darker sides of love. Shakespeare's Othello asked to be remembered as one who loved not unwisely, but too well. Is it possible to love or be loved by our pets too well? Or is what we assume is the body language and emotion of love not really love at all?

10
LOVE, HATE, SPITE, AND JEALOUSY: EMOTIONS À LA CARTE

THE SCHERINGS' MALAMUTE, Cody, creates such a tidal wave of isolation behavior that his owners are awash in a sea of guilt and indecision. Several trainers have suggested denning to calm the troubled waters. "It sounds sensible and logical," says a nonetheless skeptical Don Schering, "But I can't bear the thought of caging Cody as if he were a common criminal. We love him too much."

But one day Bonnie Schering comes home before the rest of the family to discover that Cody has knocked over a display cabinet full of china. Among the shattered family Wedgwood lie the fragments of a commemorative plate entrusted to Bonnie's care by the regional historical society.

"I hate you, I hate you, I hate you!" screams Bonnie as she repeatedly strikes the cowering dog with her heavy purse. "How could you be so spiteful? Well, two can play that game!" Yanking Cody's beloved blanket out from under him, she rips it to shreds as the bewildered dog looks on.

Meanwhile Bonnie's sister, Peggy, and her husband, George Carlisle, sample some different but equally powerful emotions. Their English bulldog, Dickins, behaves like a prince, and both owners dote on him. When the Scherings begin complaining about their troubles with Cody, however, Peggy and George decide to practice a bit of preventive medicine in the form of a new pup to keep Dickins company, a sweet whippet appropriately named Honey, bred and raised by George's sister.

"We thought Dickins was the most incredible dog in the world—until we got Honey," marvels Peggy at the end of their first month together. "But boy, does she make old Dickins jealous. Every time I pay attention to Honey, he sticks his nose between us. But that's the only problem we've had."

Actually, that's *not* the only problem. George has been wrestling with the green-eyed monster himself. "I've never seen a dog like Honey before. She's extraordinarily loving. Maybe that's why it bothers me so much that she prefers to be with Peggy."

Imagine entering what all the critics have called the best restaurant in the world and asking for your four favorite dishes. The maître d' politely replies, "We don't serve any of those here. In fact, our chefs have never heard of them." That's exactly what happens when we subject the emotions we'll be discussing in this chapter to the scientific view. While most of us can point to incidents where we and our dogs clearly express love, hate, spite, or jealousy, behaviorists rarely, if ever, acknowledge the existence of such emotions in our canine companions. Unlike dominance, submission, fear, aggression, or isolation behaviors, where we can use objective scientific data to reinforce or alter our beliefs, we're pretty much on our own when it comes to our interpretation and use of love and its many manifestations, including its troubling counterstates.

Does this mean that we dog lovers and our people-loving pets should knuckle under and remove these emotions from our relationships? Why bother having a dog at all if it's actually incapable of recognizing and sharing love? The answer is clear: Regardless of any scientific evidence to the contrary, we know in our hearts that love-hate emotions exist and exert powerful effects on our relationships. Therefore, when we enter the restaurant only to be told, "We don't know how to make that dish," we must ask to use the kitchen ouselves. It won't take us long to teach those scientific chefs a thing or two about gourmet cuisine—provided we don't blow up the stove in the process.

GOOD BITCH, BAD BITCH

Given the lack of objective guidelines with which to define loving, hateful, spiteful, or jealous behavior in our relationships with our dogs, we often assign multiple or even no definitions to these states. Nonetheless, we all recognize that these emotions are real. Bonnie Schering simply shrugs her shoulders and says, "Don't ask me how I know Cody loves us as much as we love him, I just do." Some owners feel that their dogs love them because they gobble up every morsel of food placed before them; and others believe their pets love them because they sleep on their beds. Still others know their canine companions love them because they'll attack anything that comes into their yards or, by contrast, because they wouldn't hurt a fly.

The negative sides of love—hate, spite, and jealousy—suffer similar subjective fates. Many owners recognize certain nebulous "looks" as proof of these emotions in their pets: "I could tell by the look in Dickins's eyes that he was jealous of the pup," declares George Carlisle with conviction. For many owners all the negative displays the behaviorists attribute to isolation signal their pets' jealous, spiteful, or hateful natures. Similarly owners recognize certain behaviors in themselves as loving, jealous, hateful, or spiteful. Because these feelings and their expressions are so real to us, it never dawns on us that our dogs might have no inkling of what we're feeling or displaying with our own body language. When we take this stew of definitions and expressions and liberally season it with the inconsistency that's characteristic of so many human/canine relationships, no wonder these emotions precipitate and complicate a wide spectrum of problem behavior.

Initially owners may classify a canine behavior as either good or bad, but we soon assign all sorts of motivating emotions to that behavior and then respond to it with our own emotions and body-language expressions. The loop can repeat itself endlessly, but at no point does it allow us to define or solve any problems. For example, observing Honey spit out her specially prepared treat in favor of her boring kibble, George Carlisle admonishes her, "Bad Honey, don't you love Daddy?" George sighs heavily, sinks sadly into a chair near the dog, and tries to think of other treats Honey might find more appealing. The actual problem—that George is equating love to a behavior that carries no such association for his dog—doesn't even enter his mind.

Consequently, before we can even begin to recognize problems, we must determine the linkages we make between normal behaviors and emotions. Once we recognize these specific connections, we can decide whether they benefit or harm the relationship. Finally we can decide whether we want to make any changes. When Honey has pups and protects them so aggressively she even snaps at George, should he interpret that as normal loving behavior for a nursing bitch, or should he believe the dog is behaving hatefully toward him? Ironically, while George can accept her as loving when she snaps at others to protect her offspring, he can't accept her snapping at him as anything but a mean and hateful act.

While such subjectivity can obviously cause problems for dog and owner alike, it can also prove beneficial to owners when it permits us to create and reinforce specific definitions and behaviors that fit our needs. If George believes Honey expresses her love for him every time she eats a particular food, he has concrete evidence of her love whenever this act occurs. By the same token, owners who associate specific behaviors with jealousy or spite obtain evidence of their pet's feelings whenever those behaviors occur. In such a way a relationship can grow more intimate and animated, adding a zest the behaviorists' view denies.

Given the reality and subjectivity of these emotions, let's examine each one separately to see if we can develop some basic concepts that will provide objective reference points as we seek to integrate or eliminate these emotions from our interactions with our dogs.

HOW MUCH DO I LOVE THEE?

The romantic within me longs to save love for last, proposing it both as a universal solvent to dissolve hate, spite, and jealousy and as a miraculous glue that can bind even the shakiest of relationships together. Unfortunately, the realist within me knows that, far from being the antithesis of these painful emotions and conditions, love is their very source. If we were indifferent to our pets or they to us, if we didn't want to share magnificent positive feelings with them, we wouldn't find ourselves occasionally plunged into the highly negative throes of hate, spite, or jealousy when our loving feelings are thwarted. Negative feelings can exist only when we care a great deal, whether we call that caring love or not. Therefore, if we can discover what it means to share love with our dogs, we can enhance our understanding of these other emotions too.

What does it mean to share love with a dog? Most owners recognize that it takes two to love. Moreover, both human and canine feelings and expressions of love can be classified as acts of commission or omission. We express love to our pets and evaluate their loving response in terms of specific deeds they perform or fail to perform. For example, when my dogs stop what they're doing and greet me when I enter their presence, I view their attentiveness as an expression of love. I also consider their obedience an expression of love—when, despite their desire to do so, they don't take off down the road after that noisy red pickup simply because I told them not to. I feel I express my love for them in the way I touch and talk to them and in the opportunities I create for us to share experiences together. I also feel I manifest my love when I don't get angry or impatient with them when they misbehave.

Each of us associates his or her own set of quite specific as well as nebulous body-language displays with love. Before reading on, take a moment to review those you use with your dog and vice versa. Although there are as many singular behaviors as there are owners, most agree on a few basic concepts. By understanding these we can expand our awareness of love beyond the manifestation of an idiosyncratic list of dos and/or don'ts.

Think about your own body language of love. Do you suffer from the Saint Francis syndrome? Many times owners see their dogs as poor dumb creatures who couldn't possibly survive without their intervention and benevolence. Without them, the dog is nothing. Surprisingly, the Saint Francis syndrome afflicts both dependent and independent owners. Dependent owners treat their pets as though they're too physically, mentally, or emotionally weak to survive in this cruel world; after all, what three-year-old child could? They tend to adopt an instructor's role, teaching their dogs to be human by treating them like humans, so that they can survive in a human world. Essentially they're saying the dog is an alien species, but a delightful one that, when disguised and treated as a child, can sneak into human society.

Compare this attitude to that of the independent owner, who sees the dog as an alien species of a different ilk. Such owners believe their benevolent training and companionship permits a wild creature to coexist with people; otherise the dog would be doomed to an inferior, wild existence.

In both cases the owners manifest one of the most obvious characteristics of the Saint Francis syndrome: a belief that the dog—and only the dog—experiences a lesser existence outside the relationship. Of course, the real Saint Francis never espoused such a view; he recognized the beauty of dominion. Unfortunately, when we lack confidence in our dogs' *desire* to love and be with us as much as we want to love and be with them, we feel we must dominate them in some way. Dependent owners seek to dominate their pets psychologically whereas independent ones try to dominate their dogs physically.

However, while the impartial behaviorist may see these human body-language displays as simply expressing dominance over the dog in two different ways, more likely than not the owners see their behaviors as evidence of their great love.

For example, Honey Carlisle might be the epitome of a dependent dog. Like so many dependent relationships, the one between George and Honey springs from emotionally fertile circumstances: Honey was a Valentine's Day present to George and Peggy and from his favorite sister, who then sold her kennel and grooming business and left the area. Like many dependent owners, George uses Honey to fill a void in his life. Consequently his definition of love is quite homeocentric: He cooks special treats for her, provides a special bed and toys for her, and takes her special places at special times. All of these displays add up to George's definition of love. To show her love for her master, Honey need only accept all his ministrations enthusiastically.

This dependent approach to love assaults the dog on two fronts, one physical and one behavioral. Because of the connection these owners make between (human) food and love, their pets are prime candidates for obesity, malnutrition, and a host of other medical problems related to inadequate diets. When such problems occur (and they inevitably do), the dogs are extremely difficult to treat successfully because of their owner-induced finicky and often exotic eating habits. Secondly, such pampering almost always produces a pet displaying all the negative signs of dependency. While their owners will usually attribute such displays to the dog's great devotion and love, most other people find the behaviors obnoxious.

The Scherings create a different definition of love for their wolf look-alike, Cody the malamute. To earn Don and Bonnie's love or to express his love for them, Cody need only do what they want him to do. The drawback of this particular orientation lies in the owners' assumption that they will always clearly convey their desires in ways their pets can comprehend. While this is an ideal all owners strive to achieve, it's presumptuous to assume that it automatically and invariably occurs.

While the Scherings' definition of love may seem much simpler than George's complex regime, both suffer from the same deficit: Neither orientation defines love beyond that which the dog can (and often *must*) do for and with the owner. In essence such owners say, "Because I love you, I want you to eat this (sleep here, do this, don't do that, and so on). And if you love me, you'll do (or not do) those things." Consequently if the dog doesn't perform properly or chooses to do something else, by the owners' own definitions, they have not only a misbehaving dog but an unloving one.

TELLTALE SIGNS OF LOVE AND HAPPINESS

Among the wide range of specific body-language displays owners associate with canine love and happiness, three of the most common are tail wagging, licking, and jumping. Not surprisingly all three displays serve quite practical and unemotional purposes in the wild animal.

Earlier we noted how the position of the tail provides a valuable clue to how the dog views its position relative to its environment. Tail up signals dominance, tail tucked indicates submission. We also know that dogs' eyes are much more sensitive to motion than to detail or color; therefore a waving (wagging) tail is a highly visible display. To further facilitate recognition, primitive canids like wolves sported bushy tails that, when curved upward and waved in greeting, exposed lighter-colored fur on the underside—a distinctly different visual pattern than that presented by the rigid tail of the defensive animal or the tucked tail of the submissive one. However, while the wild dog's tail serves as an effective means of communication, both the appearance and function of domestic canine tails have undergone some major changes.

Although the vast majority of dog owners point to the wagging tail as evidence of canine affection, tail-wagging tendencies actually vary greatly from breed to breed. A brief review of any book on breed standards quickly reveals that the cues signaled by the wild dog and the love and happiness so many of us associate with tails often carry entirely different meanings to members of the dog fancy. To pass muster, individuals belonging to various breeds must carry their tails low, high, straight, curled, or tucked. Some must have plumes, others cannot; some must have tails of a specific length, others none at all. Dogs bred for a particular function are often assigned tail-language to complement that function. Some hunting dogs, such as the spaniels, are expected to wag their tails constantly while pursuing quarry, then hold them rigid to signal they've located their prey. Sled dogs should hold their tails high when working, while herding canines should hold theirs low. While this may seem capricious, it makes good sense. The high tails of the sled team signal that all dogs are attentive to the driver and provide instant clues to impending intercanine problems. Because herding dogs use nips and the motion of their entire bodies to move the herd, an erect, waving tail perpendicular to the canine line of motion distracts from rather than enhances the message the dog wants to convey. (Imagine someone chasing you toward a gate while simultaneously waving a white flag behind his back; the addition of the flag only creates confusion.)

From this we can see how those interested in breeding animals for show or work would tend to select animals with strong tendencies toward these characteristic displays. Consequently, if Gretchen the German short-haired pointer comes from one of those lines that shows enthusiasm for the hunt with "violent" wagging, we can expect her to wag much more readily than Sydney the border collie, whose ancestors include some of the finest herding dogs in the country. Put these dogs in domestic situations and mismatch them with their owners' expectations and we might quickly find owners faulting pets for being overly demonstrative or too reserved. Clearly we need to learn to read our dogs as representative of a breed as well as individuals.

Earlier we also noted how animals greet and identify each other by sniffing the moist membranes where odors are most pungent, particularly around the mouth, rectum, and urogenital areas. Licking is usually a part of this ritual; the more dominant animal licks its own nose and sometimes the other's nose; and the more submissive one may nervously flick its own tongue during the process. Dogs also lick these same areas as part of their personal grooming pattern and are likewise drawn to any unusual discharges, such as those from infected ears or sores on themselves or others. As already noted, bitches will also lick their pups to stimulate them to urinate and defecate.

How we respond to being licked by our dogs depends on our interpretation of these body-language expressions. Those who view licking as a natural canine greeting respond differently from those who see dog tongues as ever-ready canine toilet paper. People licked by dogs may be delighted and exclaim, "He kissed me!" or may flee to the nearest bathroom to retch and rinse off the disgusting canine saliva and, hopefully, its host of microscopic evils.

Because licking means so many different things to dogs and people, we should avoid linking it with something as specific as love. A couple who adored being licked by their dog was appalled when the dog would literally soak their child with its incessant lapping. They didn't realize they'd established a pattern of intense reinforcement that exaggerated its meaning to the dog. Because the infant couldn't get up and walk away like the adults, the dog could lick, and lick—and lick.

It's also very difficult to teach a dog to "kiss" only one person. Having been on the receiving end more than once, I can tell you that this display can be pretty unnerving even to a veterinarian. I once entered an exam room and was met by a lunging shepherd who planted a wet lap on my lips. Hardly a threatening display in retrospect, but in that instant, when seventy pounds of dog was flying at my face, my thoughts didn't naturally flow toward happiness or love. "She's just saying hello," gushed the owner happily as I struggled to pull my heart from my throat and my stomach from my toes.

On the other hand, responding to the dog's tongue as some sort of lewd sensor dripping with vile microorganisms serves no beneficial purpose either. Wise owners adopt a more unemotional approach, seeing the tongue as an exquisitely designed, multifunctional bit of canine anatomy. This doesn't mean we can't enjoy a lapping kiss at a particular time or that we should never associate it with our pet's love. It does mean we don't *have* to be kissed to know our dogs love us. Conversely, if we find licking disgusting, viewing it as an unemotional canine function enables us to excuse our pets if they inadvertently respond to us as if we were one of their own kind.

Is it possible to stop a dog from licking people? It is, and owners can use many different methods to prevent or extinguish the unwanted display. The problem crops up frequently in young pups, whose general mouthiness often accompanies teething. When it occurs in adult animals, it's usually been accepted and even reinforced. In both cases the dog's natural body-language display may be viewed by a given owner as negative because of its focus; but keep in mind that the behavior in and of itself isn't bad.

Whenever we dislike a dog's licking, we may choose to interpret the display in one of two ways: Either the dog is doing something wrong to irritate us, or it's doing something it considers right but we don't agree with that interpretation. If we accept the former view, some sort of punishment would seem to be in order. However, if we view the problem as a normal display with an inappropriate focus, punishment makes little sense.

Given the nature of most licking behavior, I prefer treating mouthiness passively. If these displays occur at specific times—such as part of an exuberant welcome—either ignore the dog during those times or give it more acceptable ways to manifest its mouthiness, such as carrying or fetching a ball. If you don't have time to abort the display, gently but firmly grasp tongue and lower jaw while cheerfully and sincerely noting, "For me? Oh Dickens, how wonderful. I always wanted a bulldog tongue for my very own." This inevitably confuses the dog, who backs away and quickly decides this game the owner finds so much fun isn't so great and it's a good idea to keep one's tongue safely in one's mouth whenever people come around. If you find the idea of sticking your hand in your dog's mouth distasteful, wear a glove. If you're afraid to stick your hand in, don't even try it—this technique is for lickers not biters, for lovers not fighters.

The final body-language display many owners associate with canine love is jumping. Some owners will even endure scratches, bruises, torn clothing, and malevolent feelings about their pets rather than extinguish the behavior. When Cody leaps through the air, is he really expressing love? Recall how dogs signal dominance by establishing direct eye contact and placing their front paws on the shoulders of another. A more dominant young pup will exhibit these displays when it first enters its human pack because it naturally wants to discover its proper position. So when Cody jumps up on his owners and they praise and pet him because they associate the display with his love and happiness, they simultaneously encourage his dominance. While the pup is establishing its place in the pack, the owners are attempting to train it. Intuitively most people first concentrate on getting the pup to pay attention, usually by responding to its name and establishing direct eye contact with the owner. In such a way eye contact comes to be associated with attention as well as dominance.

If the dog is stable and small, the jumping may never become a problem, and the owner can maintain the love-leap link quite easily. If the dog is quite dominant and/or becomes physically large, though, problems inevitably arise. One owner took pride in her Great Dane's willingness to "dance" with her. Her feelings changed when the dog began growling when she pushed him down, and they ended abruptly when he pinned her against her car and had to be pulled off by a beefy neighbor. More typically the reinforced happy leaps become an ambivalent part of owner/canine interaction. When we're wearing old clothes and aren't carrying expensive crystal, we relish the close interaction; when we're dressed formally or carying bags of groceries or a carefully prepared report, the same display infuriates us. Some of our guests adore and encourage the behavior; others despise it and chide us for tolerating such unruly beasts. And while the instinctive canine behavior expresses dominance, such an expression within the human pack becomes more of an attention-getting display; whether the person targeted for the dog's jump likes or dislikes the behavior, the dog invariably attracts the person's attention, which, of course, perpetuates the behavior.

Consequently, it's often much easier to divest ourselves of the love-leap association and extinguish the behavior than try to train the dog to distinguish between "right" and "wrong" jumping situations. As we already noted, we can extinguish this behavior simply by ignoring it and encouraging everyone else who comes into contact with the dog to do likewise. Because many jumping canines are seeking attention, yelling, smacking, kneeing, or stepping on dog toes is not only ineffective, in some cases it actually perpetuates the behavior because of the contact established during the exchange. Dogs who are ignored invariably become bored and go lie down somewhere. Once this occurs, go to the animal, quickly squat down (so there's no need for it to jump to establish eye contact), and softly praise it. Owners of bigger dogs often discover it's easier consistently to ignore the leaping canine if they change into old clothes before entering the house and leave any packages outdoors until they're sure the dog has settled. The way they're free to concentrate on ignoring the dog—and it can require a great deal of

concentration to ignore an airborne Airedale or retriever—
rather than worrying about their clothing or belongings.

Owners who effectively abolish the leaping-love link without guilt invariably delight in the results. In addition to having
a pet that greets them and others with calm but nonetheless
genuine affection, they often reap other rewards. By decreasing
the intensity (not the quality) of our interactions with our pets,
we eliminate the stress related to the animal's memory of emotional partings and anticipation of emotional reunions; consequently these animals exhibit less negative isolation- and
frustration-induced behaviors. Owners of submissive pups or
adults who urinate submissively when exuberantly greeted or
praised discover this behavior almost always disappears when
they use more subtle body-language displays to signal love and
affection.

Tail wagging, licking, and leaping are all unemotional
body-language displays that evolved for specific reasons. If the
linkup between any of these and love works well for you and
your dog, by all means don't change it. But if it doesn't, don't
hesitate to dissolve the association and create a new one. One of
the most loving dogs I ever encountered did nothing more than
press his huge nose into one's palm and stare intently into one's
eyes for a scant five seconds or so, yet no one who experienced
the display could doubt his great love—especially not his bed-
ridden owner.

HATE AND SPITE: LOVE TURNED UPSIDE-DOWN

If we define something as critical as love in a lop- or
human-sided manner, we shouldn't be surprised when our dogs
don't fit into our definitions; nor should it surprise us when we
take their failure to do so quite personally and emotionally. Yet
we are almost always shocked when such deviations occur. Don
Schering may unemotionally justify Cody's failure to respond to
commands to come at inconsequential times by saying, "He
probably didn't hear me," or "He was distracted by the kids. No
big deal." But when Cody ignores his command to come when

Don sees his dog straying into the path of an onrushing car, Don goes berserk: "You spiteful brute, get over here this instant!" Similarly, when Cody upends an old orange crate in the Scherings' absence, Don ignores the event while Bonnie inadvertently and subliminally reinforces it: "Cody loves us so much, he can't bear it when we leave him alone." When Cody destroys a priceless heirloom, though, the Scherings suddenly see all the energy they previously saw as manifesting Cody's love flipped upside-down, now evidencing Cody's hateful nature. While a calmer, more rational Bonnie may later recognize her own feelings of inadequacy and fear as major contributing factors to her angry outburst, in the instant she spies the smashed commemorative plate and the cheerful, guilt-free, and therefore unrepentant canine, she sees red and responds to the dog's perceived hatefulness with an equal amount of her own.

George gets into a similar dilemma with Honey on a more regular basis. Since most of us consider hate such a negative and even "wrong" emotion, we often prefer to ascribe spite to lesser misbehaviors. While the Scherings and Cody experience long intervals where Cody knowingly or unknowingly adheres to his owners' definition of love with periodic plunges into the dark realm of hateful displays, Honey eats away at George's definition with all the persistence of acid rain on a marble tombstone. Were George to construe all her infractions as hate-induced, he could hardly justify maintaining the relationship. By defining the behaviors as spiteful, he can see them as expressions of love gone slightly astray. By seeing Honey's failure to eat, sleep, or otherwise peform certain tasks in a specific way as evidence of her desire to get even with him, George invents the otherwise missing dog-half of his definition.

Given their limited definitions of love, we can understand how both the Scherings and George get into trouble with their relationships. Because they can only recognize love in their own terms, the dog's failure to comply with those terms automatically indicates hate or its more socially acceptable variation, spite. Unfortunately, it never dawns on these owners that not only their pets but other people may not share their definitions of love.

Let's return to the world's greatest restaurant where you've talked your way into the fully appointed kitchen and have begun preparing your perfect meal. You select an appropriately sized saucepan, put it on a burner, and set the heat on high. Suddenly, the chef charges into the room and angrily snatches the pan from the glowing element. "You fool!" he roars. "That's my best pan—how dare you abuse it that way!" His look tells you immediately how deeply you've hurt him with what he considers your blatant disregard for his equipment. Suppose you attempt to apologize, explaining you know nothing about fancy cookware and simply chose the pan you believed best suited to your needs, but he becomes even angrier. Not only won't he believe your excuse, he accuses you of deliberately choosing that particular pan and subjecting it to high heat precisely *because* it was one of his favorites.

Of course you would feel angry and confused, especially because this person you respect shows no inclination to forgive you. As far as you're concerned, you did nothing deliberately wrong: Your only sin is ignorance. So it is with Cody and Honey; because their owners haven't communicated their love in a way their dogs can grasp, the Scherings and George shouldn't condemn their pets for not adhering to their definitions. And yet like so many owners, they do.

Whenever such incongruency exists, spite and hate will infect the relationship. Furthermore, like all emotions we experience in our interactions with our pets, these feelings feed on themselves. When Bonnie defines Cody's smashing the treasured plate as hateful, she responds to him in a hateful way: He hurt her, so she wants to hurt him and even the score. Although Honey's infractions aren't nearly so dramatic, George does respond spitefully to what he considers her spiteful behavior. If she refuses to accept his love offerings he gets even with her by ignoring her imploring looks to be held, play ball, or go for a walk.

Such owners essentially use their dogs as little mirrors, first projecting their own definitions of love, then of spite and hate, on to their pets. The dogs, in turn, reflect back on the owner, who responds with more feelings. Thus an acknowledged loving act can bounce back and forth between owner and dog many times, obviously fortifying the relationship. However, when owners view the precipitating display as hateful or spiteful, the same process ensues, and there's no way they can escape being tarred with the same negative emotional brush as the dog. While there do exist dominant, aggressive, or otherwise threatening dogs one could naïvely define as "hateful" rather than the more correct "frightening to me," the emotions of hate and spite owners commonly recognize in their dogs spring solely from their own definitions and not from the dog's behavior. Most of what we view as spiteful or hateful behavior simply results from a lack of proper training. When we accuse our dogs of being motivated by hateful emotions, we're really saying, "How could you not know what *I* want even though I never consistently communicated this to you?" Furthermore most of us consciously or subconsciously add, "If you *really* love me, you would know—I shouldn't *have* to tell you." Then by our own definitions, when our dogs don't do what we want when we want, they don't love us—and therefore, and again by our own definitions, they must hate us.

BREAKING THE LOVE-HATE-SPITE CYCLE

If we accept that hate and spite in our relationships spring from our definitions of love, we can't hope to improve the situation unless we alter our definitions. But how do we go about changing something so deep-seated and subjective as our beliefs about love? To begin this difficult task, I usually advise clients to start by evaluating their definitions in terms of the ever-present four options. Accepting all three emotions as part of their interactions with their pets doesn't appeal to George or the Scherings. Although they vigorously embrace the concept of love as part of their relationship, they vehemently reject spite and hate as "normal" components of the ideal bond they wish to forge with their pet. Not all owners would agree. Some actually count on their pet's hateful and spiteful displays and their own responses to them to validate their relationships. Surely we all know independent owners who crow, "I guess I showed Satan who's boss!" after they've won a battle royal with a dog displaying what they consider deliberately hateful or spiteful behavior. Or what about dependently oriented Aunt Tilly, whose conversation about her miniature schnauzer is peppered with anecdotes about what the dog does to get even with her for leaving him alone, changing his food, moving the furniture, having dinner guests, and so forth *ad nauseum*? Both types of owners have so fully integrated spite and hate into their interactions with their pets that such displays form the keystone of the relationships. Without these periodic negative but validating upheavals, such liaisons would have little meaning.

What about accepting spite and hate in our dogs but changing our own feelings, that is, responding positively or at least neutrally rather than hatefully and spitefully to these canine displays? Again George and the Scherings reject this option as inconsistent with their idea of a good relationship with their pet. Yet many owners whose pets display isolation misbehaviors do exercise this option. Every time they leave the dog alone, it upsets the trash or sleeps on the bed. Because they've told the dog time and time again not to do so and have punished it numerous times (albeit after the fact and therefore ineffectually), they conclude the dog *wants* to spite them. If the destruction is sufficiently minor or changing the behavior seems inordinately difficult or impossible, these owners find it easier simply to grin and bear it. As they clean up the trash or brush dog hairs from the bed, they note, "Bucky's been spiteful again" with the same lack of emotion they might note, "It's raining again." If they accept spitefulness as just as normal for Bucky as his floppy ears, it carries little negative charge and consequently doesn't adversely affect the relationship.

Can't we do away with spite and hate in our relationships simply by getting rid of those behaviors we define as spite- and hate-induced? This, too, makes sense in some situations. Earlier, I mentioned a miniature poodle, Suki, who only chewed the jackets of his master's prized record collection and only when left alone. The best permanent solution to this problem involved change. The owner could find a poodle-proof location for his records or he could train his pet not to chew them. While these choices seem obvious and simple, they can elude the owner who agonizes over giving in to a spiteful dog.

"Why should I move my things out of *his* way?" this particular owner complained. "Why should I spend a lot of time training him not to do something he already *knows* he shouldn't do?" While both objections may sound legitimate, the fact is that both focus on Suki's spite and not the chewed record jackets as the problem

Such an inability to define the problem in a way that produces solutions poses the primary stumbling block on the road to change. Defining the problem as the dog's spiteful personality makes canine psychoanalysis, mind-altering drugs, or psychosurgery the only solutions. But if we see the seemingly spite-based body-language display itself as the problem, all kinds of options become available. Sometimes we have to answer the no-nonsense question I put to Suki's owner: "What's more important to you, hanging on to your belief that your dog is spiteful or doing away with the negative behavior?"

My client opted to move his records, even though he wasn't totally convinced this simple act would "cure" Suki's spitefulness. It did, though, because the dog never did anything else to earn himself that label. In the owner's mind, chewing and spite were synonymous; when the chewing disappeared, the spite vanished, too.

Similarly the Scherings discover that all their feelings about Cody's spiteful and hateful nature arise directly from his isolation behavior and his failure to come when called. By initiating a program that combines denning with consistent daily obedience training, they remove spite and hate from their relationship in a matter of months. Like many owners, they come to view denning, which they previously perceived as imprisonment, as an extremely loving gesture. Not only does it eliminate the tension that resulted in Cody's former negative displays, it also eliminates all the negative human emotions that attended them.

George's relationship with Honey encompasses so many spite-related behaviors that it makes more sense to eliminate

the emotion than to try to do away with all the displays. However, to do this George must change his definition of love.
"That's impossible," he insists. "Love is love. You can't define,
much less redefine it." Yes and no. Love *is* love, but it's also the
result of our definitions; if we possess the self-confidence simply
to forget about rigid definitions, we can eliminate the agony we
experience when our dogs don't fulfill our love criteria.

George's list of behaviors Honey must perform to demonstrate her love includes:

- Eat her treats.
- Sleep quietly at the foot of the Carlisles' bed.
- Be obedient.
- Like what and who he likes.
- Not bark except in emergencies.
- Ignore all other dogs except Dickins.

His list of acceptable love expressions for himself requires that
he:

- Provide special treats for her.
- Walk her at specific times.
- Feed her punctually.
- Provide her with special toys.
- Play with and pet her on demand.
- Seek immediate veterinary care the instant any problem
  occurs.
- Spend as much time with her as possible.

If George perceives any deviation from these lists as unloving,
spiteful, or hateful behavior, we can see how his complex definitions of love can easily become cumbersome. Neither dog nor
owner has much leeway to change without violating some aspect of this artificial and intricate construction.

To extricate George from this emotional maze, let's pose a
simple question: If you were on a desert island without all the
treats, toys, and other paraphernalia of ownership, would you
still love your dog? "Of course I would!" claims George without
a moment's hesitation. Exactly—because love simply *is* and has

nothing to do with what we and/or our dogs do. If it does, it's because that's the way we define it, not because that's the way it has to be.

One of my favorite devices for extricating myself from the spite trap is a simple trick my husband taught me. Whenever one of my dogs does something I perceive as spiteful—lying on the rug in front of the door to trip me, for example—I ask myself, "What does it mean when Dacron lies on the rug in front of the door?" And I force myself (and it *was* necessary to force myself at first) to answer, "It means Dacron is lying on the rug in front of the door." In other words, it doesn't mean he hates me; it doesn't mean he wants to trip me; it doesn't mean he wants to get even with me for being slow with his food or not taking him with me to the store.

By using this technique George quickly accomplishes two major changes in his relationship with Honey. First, he's able to eliminate those awful feelings that come in the wake of the negative behaviors. This enables him to evaluate these displays and make necessary changes objectively. Second, by defining love as an inseparable part of his relationship with his dog, he gains confidence from knowing that regardless how poorly either of them behaves, love is ever-present. Instead of being certain of his own or Honey's displays as evidence of a lack of love and the presence of hate or spite, he now sees them as negative behaviors that only temporarily mask the constant love between them.

While some may argue with the behaviorists that this is all a mind game, surely it's every bit as legitimate as the spite and hate we choose to see in our dogs' and our own behaviors. Because we do attribute emotions to our dogs, why not choose the most expansive interpretations, ones that allow us both the maximum freedom to experience our respective individual natures in a loving, supportive environment?

EXPELLING THE GREEN-EYED MONSTER

Of all the offshoot emotions that can grow out of love, perhaps none causes more pain than jealousy. If we feel strongly

about others but lack sufficient confidence that he or she recip-
rocates that love, jealousy can penetrate every corner of the relationship. Even if we don't accept love as a valid part of the
human/canine bond, we may still react jealously to certain
body-language signals. The green-eyed monster most often attacks when two (or more) individuals contend for another's attention or affection.

The specific body language of jealousy varies from individual to individual. For example, as soon as I lavish any attention
on one of my dogs, the other immediately sticks his or her nose
in the affair. The behaviorist in me can't attribute such gestures
strictly to an assertion of pack structure because it makes no difference whether I'm concentrating on the relatively more dominant male or the quiet submissive female—the other will
immediately trot over and attempt to position his or her body to
get equal, if not more attention. The male more frequently
overrides the female by placing his front paws on her shoulders
while simultaneously pushing her aside. The female counters by
flattening her ears and driving her body like a wedge between
her littermate and me at ground (submissive) level.

Because my dogs recognize almost all humans as dominant,
they don't interfere with interhuman displays of affection.
Other owners aren't so lucky. Many dogs don't hesitate to force
themselves between man and woman or parent and child to demand a share of the attention, a dominant behavior expressed
most commonly by dogs whose owners have knowingly or unknowingly reinforced such displays. For example, Don Schering
often plays with Cody in a manner quite similar to that used by
littermates. He grabs the malamute by the shoulders and flips
him over, "growls" and "barks" at him, and chases him about
the room. Sometimes he lets Cody chase him, and he rolls over
and wrestles with the dog as part of the game. Bonnie never
participates in such play because she thinks Cody is too big and
the game's too rough. However, when she's alone with him, she
does encourage alarm barking and similar "protective" displays
that signal her submission to the dominant dog. She also permits
him to jump up on her, drag her when they go for walks, and
muscle his way onto the couch beside her.

So when Don and Bonnie embrace, it's no wonder Cody pushes his way between them. Although his stable relationship with his owners prevents him from going beyond persistent nudging (which the Scherings find humorous at best, but only irritating at worst), the competition between other dogs and humans can become fierce enough to precipitate biting incidents. When that happens, most owners quickly redefine the once-tolerated jealous behavior as evidence of the dog's hateful and vicious nature.

Something quite different but not unusual occurs between Dickins and the Carlisles when the beautiful Honey enters the household. Previously owners and dog seemed equally, though not slavishly, devoted to one another. Unlike the delicate and more dependent whippet, Dickins is a stoic rock, a dependable but unobtrusive family member, whose personality results not only from his breeding but also from the reinforcement of such behaviors when he was a pup.

When Peggy and George select Honey, their choice reflects changes that have evolved in their own attitudes during the five years they have owned Dickins. Whereas the reserved bulldog beautifully reflected their past ideas of their solid and responsible self-image, experience has increased their confidence and mellowed their outlook; now they're not nearly so embarrassed about expressing their feelings toward a dog, and the charming Honey awakens certain dormant emotions. In no time dog and humans find themselves involved in what among humans would be called a blind romance, one that has a startling impact on the once-stoic Dickins. The once-mild-mannered companion soon turns into a seemingly disruptive and belligerent monster, constantly interrupting any interactions between his owners and the new dog.

In this situation, not only must the bulldog incorporate a new member into the pack, he must also respond to strikingly different behaviors in his owners. Even if Peggy and George had responded exactly the same way to Honey as they had to Dickins, the two dogs would still have to exchange body-language displays to establish a new pack structure that could accommodate both dogs. In fact, much of what Peggy interprets as Dickins's jealous intervention signals nothing but a display of dominance over the newcomer. When Peggy encourages the pup to jump up on her, a dominant display she doesn't permit Dickins to peform because of his size, the bulldog and Honey receive some confusingly mixed pack signals. To put the pack in order, Dickins flashes dominant cues to Honey, who responds by backing away from Peggy, flattening her ears, tucking her tail, and making other appropriate submissive gestures to indicate she knows her proper place. Peggy, though, sees the entire episode as solid proof of Dickins's jealousy.

In addition to the displays he directs toward Honey, Dickins also alters his behavior toward Peggy and George. Although his aloof attitude is a personality trait his owners chose to reinforce, it's not necessarily the *only* one Dickins is capable of expressing. Given the social nature of dogs, they prefer and enjoy close contact with other compatible pack members, but often the owner's different orientations will encourage only certain intimate interactions. Consequently the dog learns to display those behaviors the owners associate with acceptable canine personality traits. Close observation of any dog's interactions with different people quickly reveals that such displays aren't absolute. Regardless of the behavioral or emotional motivation we may assign various behaviors, most dogs behave differently depending on whether they are interacting with children or adults, men or women, their own family on strangers, companion dogs or strange dogs.

Therefore, when the Carlisles become more openly expressive with Honey, they shouldn't be surprised that Dickins's overall behavior changes. The simple fact that their behavior is unusual attracts him. Furthermore, if owners obviously express delight in a new type of behavior, surely the naturally curious canine will want at least to sample if not openly participate in this new display.

So why do we so easily see jealousy in dogs like Dickins? Like the other purely emotional states we've described, in order to project jealousy on to our pets we must first recognize it in ourselves. Jealousy differs markedly from spite or hate, which result from aberrant definitions of love; jealousy arises more from an aberrant philosophy regarding the *amount* of love available. Love is love; owner and dog needn't do anything specific to prove its existence. Yet the green-eyed monster feeds off our belief that love has strict quantitative limits. According to this view, owner and dog have a finite amount of love, which they then dole out to others; because there's only so much available, if one gets more, another must get less. Thus the addition of one more dog to a household means that the original pet can now receive only half as much love as before.

Such a limited view of love leads people to guard any love they receive from others and carefully apportion their own. Surely something so special shouldn't be wasted. Moreover and by our own definition, the addition of anyone or anything that merits love from one who also loves us, must of necessity lessen the amount of their love available to us. Peggy and George project just such beliefs on their relationships with Dickins and Honey. Because they're not confident in their ability to love Honey and Dickins differently *and* equally, Peggy and George feel guilty about their attachment to the more demonstrative Honey. She fulfills a different love-definition than they applied to Dickins five years ago: Doesn't that mean that poor Dickins subordinated his own desires for a more openly affectionate relationship to fulfill their former definition and that now they view him and that definition as inferior and even "wrong"? And what about this observation so frequently made when jealousy attacks: "After all Dickins has done for us, how can we be so insensitive and cruel to love another so much?"

While Peggy and George experience great guilt as a result of their beliefs that Dickins is jealous, George must simultaneously deal with his jealousy of Honey's attachment to Peggy. He selected and trained Dickins as he did because he wanted a "man's dog," a strong, dignified companion to mirror his ideal self-image. With success and maturity, George no longer feels his dog must be like him and, in fact, he finds himself more intrigued by dogs who aren't. Honey is so different from both himself and his canine alter ego, Dickins, that her every move fascinates him. However, George is a dominant male, the undisputed head of the Carlisle pack. Consequently there's no way

Honey will relate to him as freely and openly as she does with the less-dominant Peggy. The quicker George accepts this simple unemotional fact, the happier he'll be. Can he accept that Honey doesn't relate to or love him *less*, just *differently*? Can he accept that sometimes Honey appears to respond preferentially to Peggy's body-language cues simply because they're more attuned to Honey's needs *at that time*? Honey doesn't love George less, nor does she always prefer Peggy's company. Like Dickins, like each person in the world, Honey simply has different needs at different times—and part of loving her means that George should want and allow her to fulfill those needs in the best way possible.

Of course, such a generous attitude lies beyond our grasp if we believe that we and the dog have only so many units of specifically defined love to share. Under those conditions, if Honey lavishes affection on Peggy, she can't possibly love George simultaneously. If George and/or Peggy love Honey, they can't possibly love Dickins as much as they did before Honey joined the family; and if Honey and Dickins love each other, then neither can love George and Peggy as much as either would if they were the only dog in the household.

Keeping track of the green-eyed monster usually takes a great deal of time, and banishing it takes an incredible amount of confidence and patience. Because jealous feelings invariably damage relationships, their continued existence can only lead to change or termination.

MAKING THE CHANGE

Confidence more than anything else is necessary to vanquish the green-eyed monster. Unless we confidently accept the quality of our feelings for our dogs, we're bound to preoccupy ourselves with the quantity. Suppose you adore apple pie and dearly love your golden retriever, Bluebell, which you consider the most perfect dog in the whole world. If I ask you, "Which do you love *most*, apple pie or Bluebell?" you wouldn't hesitate to call that a stupid question because your two loves differ in quality. So, too, the relationship between every human being and every dog. That Dickins behaves differently from Honey doesn't mean he loves his owners more or less; that Honey's more submissive nature leads her to spend more time with Peggy doesn't mean she loves her mistress more. The behaviorists say that all these displays merely reflect the dog's expressions of the various unemotional behavioral states that attend any pack situation. The more subjective observers among us note that the positive and negative emotions we attach to these displays are our own creations and that only we can change them.

Why do I interpret the attempts of both my dogs to be petted at one time as signs of jealousy? Why don't I choose to see their eagerness for affection as their way of saying they believe I'm capable of loving them both equally and simultaneously? It's a choice I make based on some old beliefs—beliefs that change as my confidence grows.

However, I can change much more easily than George can because my attitude regarding jealousy doesn't involve my dogs' relationships with other human beings. Other people always complicate the change process. George might admonish himself by saying, "It's only a *dog*, for Pete's sake!" and glance about quickly lest someone overhear him. Surely only a fool would be jealous of a dog. But love can make fools of us all.

Honesty should be the watchword of any changes George initiates. Not only should he be honest with himself about his feelings but also about what those feelings are telling him about his definition of love. And because Peggy is an intimate part of his feelings, he must also talk to her honestly about them. Sometimes behaviors we believe others use to isolate us from our dogs' love completely escape their notice. When George sheepishly admits that he finds Honey's devotion to Peggy upsetting, his wife tells him she does, too. "She's darn close to becoming a pest, and I wish you *would* spend more time with her so she's not so dependent on me." This not only relieves them both, it inaugurates a plan whereby George will spend more time with the dog.

How can George resolve his jealousy if Peggy finds her relationship with Honey ideal? Because George is the source of his own feelings, regardless of what Peggy or Honey do, only he can change them. In this case George must determine whether to maintain his definition of love and the resultant jealousy it precipitates or alter his beliefs about this most powerful and intimate emotion. Before he can enjoy a stable relationship with his pet, he must recognize that the choice lies solely in his own hands.

THE TIE THAT BINDS

The bonded relationship allows maximum freedom for the expression of love. It also nurtures the inescapable paradox that governs this emotion and its satellite states—hate, spite, and jealousy. Love is simultaneously everything and nothing in the relationship. We can't hear, feel, taste, touch, or smell it, yet every interaction with our dogs signals its presence or absence.

It really doesn't matter whether I love my dogs the same way or to the same degree you love yours, or that behaviorists say dogs can't recognize or display love at all. All that matters is that we realize the intimate and inextricable connection between our own definitions of love, hate, spite, or jealousy and how we project them on to our dogs. No Supreme Canine somewhere vests each newborn pup with specific behavioral displays for communicating these emotions; only we can put them there. If we don't like what we see, only we can change it—either by changing the dog's behavior to eliminate those displays we associate with the negative emotions or choosing to view those same behaviors in a neutral or more positive light.

The highly subjective nature of these emotions provides us with some of the most frustrating and intriguing problems of pet ownership. We must know ourselves first and accept responsibility for what we choose to see in our dogs—both the good and the bad. We must see hate as a variant of love, spite as a measure of our own limitations, and jealousy as evidence of our own lack of self-confidence.

When we confront the darker side of ownership, when we deal with our feelings regarding hate, spite, and jealousy emanating from our dogs and the evocation of these same negative emotions in ourselves, we can't help but learn. The behaviorist might tell us that stripping all the hate, spite, and jealousy from the relationship returns it to its naturally unemotional state; but every bonded owner knows that once all these negative emotions are removed, invariably what's left is love.

Having dealt with the lightest and darkest emotions that affect our relationships with our pets, let's explore that uneasy, gray world of sorrow and depression.

11

SADNESS, SORROW, AND DEPRESSION: THE VALLEY OF SHADOWS

FOR MONTHS GLORIA WEINSTEIN has been planning and saving for her vacation, a reward to herself for successfully surviving a series of personal and professional crises. A two-week Caribbean cruise seems to offer the ideal respite, with one exception: She can't take Darwin, her fox terrier, with her. While this saddens her, her excitement overrides any guilt and unhappiness about kenneling him. As the dog mournfully watches her fold her bright new sundresses, she gently chides him, "Don't look so sad. I'll only be gone two weeks, and the Goodriches will take good care of you."

As the departure day draws nearer, Darwin's apparent sorrow increases. Preoccupied with last-minute errands, Gloria doesn't take time for the terrier's usual walks or daily play sessions. Feeling twinges of guilt, she vows to rectify her negligence by treating him to a long romp in his favorite park. She barely notices that Darwin hardly looks up when she takes his leash from the hook by the door; normally he perks up his ears and jumps gleefully the instant she touches it. Although he fails to lick her hand excitedly when she hooks the lead to his collar—another usual event—she's far too busy daydreaming about her cruise to notice the change in his behavior.

She *does* notice when she tugs Darwin toward the door and he refuses to budge. "Please, Darwin, I know you're going to miss me, but that's no reason to mope around." She tugs again, but the sad-eyed terrier remains crouched and motionless. "Okay, be a dope, but you're not going to make me feel bad." She unhooks the leash, replaces it on its hook, and runs her errands alone.

That evening Darwin won't eat his supper, a fact that wouldn't normally bother Gloria, except that he didn't eat his breakfast either. Perhaps last night's impromptu bon voyage party upset him? Has all her packing and running around thrown him off his schedule? "Poor little Darwin—too many changes for my poor pooch!" Once again Gloria subdues any feelings that something is wrong. "I've earned this trip, and I'm not going to let a dog's sadness spoil it for me."

The next day Gloria leaves Darwin at the kennels. "I'm sure he'll be fine after a few days," says Ms. Goodrich as she takes the terrier from Gloria's arms. Gloria gives her pet an extra hug and tries to hide her tears.

Two days later the Goodriches rush Darwin to the veterinary hospital. "I don't know if he's going to make it," says the vet, shaking his head sadly. "I suspect he's been sick for most of the week."

Unlike the other displays we've studied, those associated with sadness, sorrow, and depression can have grave medical as well as behavioral consequences for our pets. Because Gloria expects Darwin to be sad about her leaving him for two weeks, she interprets all the negative body-language cues he exhibits in light of this belief. Unfortunately, she misses vital clues signaling his deteriorating health. Is Gloria insensitive, uncaring, and stupid? Indeed not; she has merely fallen victim to a complex set of ideas that evolved during her relationship with Darwin and his littermate, Pascal, owned by her employer, Ted Crabtree.

LAYING A SHAKY FOUNDATION

Although Gloria and Ted didn't plan to purchase pups from the same litter, they both responded to an ad in the in-house newsletter at the electronics firm where they work and only discovered later that their pups are brothers. After Gloria and Darwin spend an evening with the Crabtrees and their pet, Gloria decides that however the Crabtrees raise Pascal, she'll do the opposite with her own dog.

"Aren't these just the most sensitive dogs," gushed Laurie Crabtree. "If I ignore Pascal the least little bit, he stares at me with those great big sad eyes until I pick him up and cuddle him."

Later, when Ted put the dog's food on the floor and Pascal looked at him, Ted immediately called to his wife, "Laurie, Pascal doesn't like this food. Don't we have something special for him tonight?" Ted backed up and accidently bumped the pup's leg, causing Pascal to emit a surprised yelp. "Oh, you poor, poor puppy dog, I'm so sorry. I didn't mean to hurt you. Here, let me rub it and make it all better." A short while later Pascal yelped again for no apparent reason, yet Ted repeated the process, convincing Gloria that her boss was well on his way to creating the most spoiled dog in the world.

By the end of the visit, Gloria determined to do everything in her power to keep Darwin from becoming a manipulative, demanding ingrate like his littermate. During the year following that fateful visit, Gloria sticks to her vow. While Pascal becomes more and more dependent and obnoxious, Darwin becomes more and more tolerant and well-disciplined. At times, though, Gloria has to work very hard to ignore the soulful looks that seem to plead for longer walks and prolonged play. Every time she catches herself about to give in, she remembers Pascal and steels herself: "No, I know you want more, but I'd never forgive myself if you turned out like your brother."

And so while the Crabtrees attach dire psychological and medical meanings to every stare or sigh emanating from Pascal, Gloria spends an equal amount of energy conditioning herself to ignore similar transmissions from Darwin. Throughout the process, the Crabtrees and Gloria never doubt they're pursuing the ideal path toward a positive relationship with their pet.

The fact that one year later Darwin lies nearly comatose in a veterinary hospital and Pascal is the focus of escalating arguments in the Crabtree household clearly proves that the owners' respective paths were somewhat less than perfect. While having a clear idea of what we want from our relationships with our pets directly affects the ease with which we may incorporate or eliminate any emotions and body-language expressions from our interactions, this is particularly critical when dealing with the diverse and often contradictory states of sadness, sorrow, and depression. If we're unclear about what we want from our pets and what these states mean to us and our dogs, even the most perfect path won't take us where we want to go.

What are the purposes, the foundations of the relationships, between the Crabtrees, Gloria, and their pets? First-time dog owners Laura and Ted want to do everything right, but because they lack experience and confidence, they expect to make mistakes. Consequently they're keenly sensitive to any change in Pascal's body language that might indicate they've done something wrong. Gloria, on the other hand, initially acts on a rather nebulous idea about getting a dog for companionship; then she quickly supplants this with a desire *not* to own a dog like Pascal.

From these two cases we can see how fuzzy ideas and expectations can easily get owners into problems with their pets. While the Crabtrees know they want to accomplish the mechanics of pet ownership perfectly, Gloria knows only what she doesn't want and has no clear idea of what she does. In the process of indulging Pascal's every whim in response to his sad looks, the Crabtrees create a pet they guiltily describe as a "self-centered, nasty little tyrant," a negative view friends and neighbors would heartily second. Simultaneously Gloria creates a pet whose personality and behavior differs markedly from his littermate's. While on the surface she achieves her purpose admirably, as with the Crabtrees, the result is hardly what she wanted. Can she possibly enjoy her long-awaited vacation once she receives the cable informing her of Darwin's perilous condition?

Such problems don't result from insensitive, hateful, or malicious attitudes; nor do they spring from any lack of interest in or commitment to the relationship. They occur because the owners have set something other than a mutually satisfactory relationship as their goal. Lacking the proper goal, they wind up winning some battles but losing the war.

Owners like the Crabtrees are like travelers who worry primarily about the kind of car they drive and the quality of the road; because they pay scant attention to their destination, they're often shocked when they discover where the car and road have taken them. Those like Gloria, motivated by what they don't want from a relationship, might as well claim, "I don't care where I go or how I get there as long as I don't come anywhere near the city (ocean, desert, or whatever)." When they discover that the fulfillment of this criterion doesn't guarantee a perfect holiday, they feel cheated.

To further complicate matters, owners who encounter problems as a result of misinterpreting sadness, sorrow, and depression in their dog's body-language signals often succumb to the Charlie Brown syndrome. Remember the question manager Charlie Brown invariably pondered after his baseball team lost game after game: "How can we lose when we're so sincere?" If we sincerely want to do the "right" and the "best" things for our dogs, how can we and they possibly fall into such a turbulent sea of negative and even life-threatening emotions and behaviors? How can the Crabtrees find themselves on the brink of having Pascal put to sleep at the very same veterinary hospital where Darwin fights for his life?

A MATTER OF DEFINITIONS

Not surprisingly, the behaviorists don't acknowledge either sadness or sorrow in wild-dog interactions. Nor do they acknowledge the sort of emotional depression humans generally define as a state of dejection or hopelessness. Instead they (or, more precisely, their medically oriented colleagues) recognize *physiological* depression, a decrease in an animal's ability to function physically. This state represents an intricate mechanism with which wild animals fend off disease organisms without the benefits of medications and the other, often complex accoutrements of modern medicine available to domestic pets. Depression-related immobility enables the animal to maintain the elevated body temperatures associated with successful disease conquest in two ways:

- The less the animal moves about, the more energy is available for the disease-fighting process, because it takes a great deal more energy to move than to lie still.
- The immobile animal can more efficiently conserve its critical body heat by curling up in a ball, increasing the insulating air spaces between the fur, and other heat-conservation techniques.

Furthermore, we may speculate that the depressed animal presents fellow pack members with body-language signals sufficiently close to submission that they'll leave it alone; fighting is an energy-intense activity the sick animals can ill afford.

While the distinction between emotional and physiological depression may be quite clear to the behaviorists—if for no other reason than that, by their own definition, *all* depression has a physiological cause—we owners don't fare so well. For example, suppose just as the Crabtrees are about to depart for an evening of fun, Pascal fixes them with a soulful look and vomits. Like most owners, the Crabtrees immediately deal with this behavior by wondering, "Is something wrong with the dog?" If they answer "No," they clean up the mess and depart; but if they decide "Yes," then they explore the specific nature of the ailment. How they explore it depends on their relationship with the pet *at that time*. Let's examine several scenarios:

- As the Crabtrees dress for their gala evening out, Laurie suddenly remembers it's Pascal's birthday.
- The day of the party Laurie switches Pascal to a different brand of dog food and Ted slips him more than the usual number of tidbits at dinner.
- This is the fourth time Pascal has vomited in two days.

As objective observers, we can see how the first scenario could easily lead the Crabtrees to think Pascal vomited because he's devastated about being left alone on his birthday; he vomits either in response to his own disappointment and unhappiness or to get even with his owners for abusing him. In the second situation, the Crabtrees attribute Pascal's vomiting to both his sadness and major dietary changes. But the third scenario could spell trouble: Does the terrier have a serious medical problem?

Given the Crabtrees' lack of confidence and experience, and in light of their more dependent orientation, they routinely think they've somehow failed as pet owners. In scenario one, they cruelly forgot his birthday; in scenario two, they fed him improperly; in scenario three, their inattentiveness exposed the poor defenseless creature to some wretched gastrointestinal virus. Regardless of whether they perceive the vomiting as arising strictly from Pascal's sadness or from some physical ailment, the Crabtrees blame themselves.

If the more independently oriented Gloria encountered three similar scenarios her interpretations would be quite unlike the Crabtrees'. While she guiltily and sadly recognizes she's leaving Darwin alone on his birthday, she sees his vomiting as an act of spite, an attempt to manipulate her; she smacks him and banishes him to the backyard. Because she considers feeding him anything but kibble a sign of giving in to him, she can easily dismiss any food-related vomiting unless her dinner guests slipped the dog table scraps against her wishes. It's the third scenario that creates the worst problems for owners like Gloria, who link a specific biological behavior to a negative emotion. If Gloria doesn't want to respond to Darwin because she believes he displays his displeasure by vomiting, she'll probably see his continued vomiting as a persistent attempt to manipulate her. The more he vomits, the more intensely she will ignore him. Whereas the more dependent orientation leads owners to seek physiological reasons for the perceived emotional depression relatively quickly, the independent view persuades owners to interpret the dog's deteriorating physical condition as evidence of its desire to win a battle of wills.

While the dependent orientation might seem infinitely preferable from the dog's point of view, these owners often become entrapped by their own caring definitions. Suppose by virtue of his intelligence and lack of any stabilizing pack structure, Pascal becomes one of those not uncommon uptight and fearful dogs who respond to every change in the environment with increased gastrointestinal motility. Sometimes these dogs are just gassy; other times they vomit or have diarrhea "for no reason at all." Because of the owners' desire to do the right thing, they frequently leave the veterinary clinic frustrated over their dog's clean bill of health. While occasionally the dog may be subjected to a diagnostic workup including stool samples, blood tests, and X rays, normally its general good health except for the periodic flare-ups leads most vets and owners to follow a more conservative route. Many owners like the Crabtrees come to accept these periodic digestive upheavals as the normal consequence of life with a "sensitive" dog with a "sensitive" stomach.

The subtle corrosive quality of this physiological-emotional connection lies in its ability to eat away at the owners' already deficient self-confidence. Before, they only felt responsible for their dog's sad looks and misbehaviors; now they're also responsible for its intermittent bouts of flatulence, vomiting, and diarrhea. Furthermore, depending on how the veterinarian presents his or her diagnosis, these owners could go from trotting to the vet's every other day to resisting going at all lest the doctor think them foolish. Obviously the veterinarian who dismisses the dog's medical problems as psychosomatic or "all in Phoebe's head" puts the owner in an awkward position: It's bad enough the doctor thinks Phoebe has a few loose screws; surely if her owners persist in seeing the digestive upsets as potential medical problems, they risk being viewed in the same way.

BETWEEN A ROCK AND A HARD PLACE

Differentiating between sadness and depression requires the combined skills of a canine psychiatrist, a behaviorist, and a veterinarian. Unless we understand the emotional and behavioral roots of sadness and the medical ones of depression, we can't hope to untangle these two quite opposite problem states and their mutually exclusive resolutions. Does this mean we're doomed to flip a coin—heads it's behavioral, tails it's medical—every time our dogs stop bubbling over with joy?

Most certainly not; however, it does mean we need to exercise caution when we evaluate those behaviors we associate with sadness and depression. While we've already seen that we can easily define most emotions in terms of the dog's body-language expressions and our own body-language and emotional responses, this approach doesn't lend itself to the management of sadness or depression because the time available to evaluate and resolve the related problems may be quite limited. To be sure, a vicious, biting dog certainly poses an immediate problem, but that dog can be confined in a secure place until the owners find time to define the problem and consider their options. Often when we're differentiating sadness and depression, we don't have any time to waste. When we enter the realm of body-language expressions, an incorrect emotional linkup may cost the dog its life. On the other hand, an erroneous physiological one may result in time and money wasted on unnecessary veterinary visits, as well as possibly earn you the vet's scorn for overindulging your canine hypochondriac.

Because the two very different kinds of problems resulting from these states throws us into a high-stakes game with too little experience and not enough time to make a careful choice, can we develop a new system of evaluation to handle them? If we possess neither the time nor the expertise to evaluate these types of problems, can we rely on our knowledge of what's right or normal for us and our dogs? Absolutely. This is the most practical and effective solution to such a dilemma.

Whenever we enter the arena of physical and/or medical problems, whether they elicit concurrent emotions or not, we need to know what's normal for us and our dogs. If we don't know what's normal, we can't possibly recognize abnormal even under the most ideal circumstances; and given those trying circumstances surrounding sadness and depression, knowledge of normal behavior becomes a valuable tool. Nevertheless, establishing what's normal involves more than thumbing through a veterinary text and comparing the average temperature, heart rate, and number of daily urinations and bowel movements to Phoebe's. It also involves becoming aware of our dogs' psychological norms, those body-language expressions we associate with its stable daily life.

For example, when Gloria reflects on the behavior of the happy, sad, or ill Darwin, she uses the following descriptive words and phrases:

Happy	*Sad*	*Ill*
Alert	Quiet	Vomits
Bouncy	Listless	Coughs
Noisy	Sighs	Has diarrhea
Cheerful	Stares	Won't eat or drink
Impatient	Off-feed	Has hot nose

Although Gloria considers Darwin a normally happy dog, she differentiates between his exuberant happy display when she enters the house and the behavior that occurs when they're simply enjoying each other's company. And that's the way she wants it to be. Imagine living with a dog that always shows the same degree of bounce and enthusiasm. At day's end both dog and owner would be exhausted, and the owner could easily come close to feeling guilty and frustrated. Obviously, *normal* and *happy* aren't synonyms, as Gloria's description of the "normal" Darwin reveals:

Normal

Low-keyed

Well-behaved

Patient

Eats when hungry

Responsive but not pushy

As we compare this list with the other three lists, we see some clear distinctions between the happy, sad, and ill Darwin, but we also see a rather fuzzy line between "sad" and "normal."

When does "eats when hungry" give way to "off-feed"? What makes the normally low-keyed Darwin different from the quiet one? When does Darwin's normal responsive gaze become a sad stare?

The bridge linking these two lists, one normal and one abnormal, is Gloria's emotional interpretation of the particular event. If she believes Darwin has reason to be sad, she's more likely to perceive the terrier's gaze as morose. If she believes he should be perfectly content to lie next to her, she relishes his low-keyed personality; and if she believes he'd much rather be playing ball in the park, she sees his immobility as evidence of deep sadness.

While such a hazy distinction between the psychologically normal and the sad Darwin may balance itself over time (Gloria misinterprets his sadness as normal or vice versa with equal frequency), the vagueness causes more problems when she fails to relate similar body-language signals to medical depression. Such displays comprise an early-warning system signaling impending physical malfunction. Given Gloria's vague distinctions, she can't tell when Darwin's physically ill until he shows unmistakable signs, such as coughing, vomiting, or diarrhea, and often protracted signs at that. The signs of depressed biological function that veterinarians associate with the early stages of many diseases—loss of appetite, reluctance to move, loss of interest in the surroundings—parallel Gloria's description of sadness so closely that they're indistinguishable. Thus she misses the body-language cues Darwin flashes to signal his illness.

We can expect more dependent and less confident owners like the Crabtrees to develop even more complex and highly subjective views of their dog's emotional states than independent owners. In addition to those parameters Gloria associates with Darwin's sadness, the Crabtrees include Pascal's sneezing, gagging, vomiting, growling, nipping, and scratching, among others. Their list of illness signals also lengthens to include many of their sadness signs as well as a wide variety of discharges, limps, "looks," and variations in color, texture, and temperature of all visible terrier parts. Because their lists of abnormal conditions are so long, they loosely define the "normal"

Pascal as "everything else"; and because they're left with so few signs not associated with some negative emotional or physical state, they rarely describe Pascal's behavior as normal. Although his owners initially wanted to maintain a continually positive and perfect relationship, they soon find themselves swinging through a manic-depressive pattern that eventually leads them to fear that the dog and/or they are somehow basically defective.

ESTABLISHING THE NORM:
EVOLVING WITH DARWIN

To extricate herself and Darwin from the troublesome aspects of their relationship, Gloria must clarify what she expects from it, defining her goals in more meaningful terms and paying stricter attention to what's normal for her and for Darwin. Only then can she hope to create a solid bond. If she redefines her goal as "developing a good relationship with Darwin" rather than owning a pet unlike his obnoxious littermate, she automatically generates more opportunities. But even this positive first step won't produce results unless Gloria recognizes what's normal for them both. Because her descriptions of his normal sadness look almost exactly like those associated with medical depression, she'll undoubtedly get into trouble whenever her terrier becomes ill. Therefore she needs to concentrate on learning about the signs of canine depression.

Initially Gloria succumbs to the deadliest emotion shared by many owners who mistake critical symptoms of illness for canine sadness: guilt. When Gloria disembarks from the cruise ship, she can think of nothing but her pet. Is he dead? Is he alive? Will he be an invalid? Even before she unpacks, she calls the veterinary hospital, then sobs with relief at the vet's assurances that Darwin will recover: "It was touch and go for a while, but he's a tough little dog with all the spunk of a true terrier."

When Gloria retrieves her dog two days later, she asks the veterinarian to explain what happened and what, if anything, she did wrong. As he describes the early-warning signs, Gloria kicks herself for attributing sadness to symptoms the doctor associates with illness. Lacking any knowledge of normal canine temperature or hydration, she never noticed the fever and dehydration. The veterinarian shows her how to insert a rectal thermometer and read it: "See where it says 102 degrees Fahrenheit? That's normal for Darwin, but some healthy dogs may have temperatures up to 102.5 or as low as 101.5. When we admitted Darwin, his temperature was 104.5, very high for any dog and a sure sign that something's wrong."

In response to her questions about dehydration, the vet tells her that animals with fevers often shun food or water, even though the fever creates greater demands for both. Small dogs like Darwin, particularly, can quickly become seriously dehydrated. The vet shows her how to gently pull up a fold of skin along Darwin's side: "Normal, well-hydrated skin is quite elastic, so that when you let go, it pops right back into place." The more dehydrated the animal, the longer it takes for the skin to return to its normal position. Of course, different breeds and individuals respond differently, which is why it's so important that each owner know what's *normal* for his or her pet. The skin response of baggy bulldogs, sleek salukis, a young pup, and a geriatric canine with less-than-perfect kidneys each provides their owners with different norms.

Again, if we don't know what's normal, we can't possibly know what's wrong. And if we link signs of physical malfunction with negative emotions, we lose this valuable early-warning system. Imagine how devastated Gloria would feel if she chose to hang on to her false view of these body-language expressions and her subsequent guilt about their consequences. Owner and pet could easily become involved in a deteriorating cycle wherein Gloria's misinterpretations plunge her into guilt, which in turn causes her to see more sadness in Darwin's expressions, which then intensifies her guilt even more. Because neither guilt nor Gloria's definition of Darwin's sadness serves any useful purpose, doesn't it make sense simply to eliminate these two emotions from the relationship?

Deciding to do so, Gloria initiates the process of recognizing the normal Darwin. Every time she catches herself defining a particular look as "sad," she objectively observes the terrier's behavior: "What is he doing that makes me think he's sad?" Then she analyzes her answer. At first her reasons seem derived from her own inadequacies—her guilt regarding something she did or didn't do for or with her dog. However, she forces herself to recognize this as an unsubstantiated reason, one that can only hinder her relationship with Darwin.

Within several weeks Gloria recognizes that Darwin's normal behavior includes a particular pattern of eating, drinking, and elimination, the way he stretches out beside her, and a series of expressions she redefines as resting or contemplative rather than sad. Ironically, a high point during this period occurs when Darwin comes down with a mild infection. Because she's able to read the body-language signals objectively rather than emotionally, she recognizes and treats the problem early and effectively. Not only doesn't she suffer any guilt, she takes joy from knowing that her new awareness enables her to achieve her goal—a solid relationship with a healthy pet.

EASING THE PRESSURE ON PASCAL

When the Crabtrees hear about Darwin's dire illness, they feel guilty for an entirely different reason: They catch themselves thinking how nice it would be if Pascal came down with something that would get him out of the house for a few days— or even longer. "A dog this diabolical never gets sick," mutters Ted as he simmers yet another experimental meal for the finicky terrier: beef tenderloin tips with ground calves' liver. "No self-respecting germ would come within a mile of him!" Can Ted possibly be talking about that ideal little terrier the family dedicated themselves to creating less than a year ago? Unfortunately, yes.

While Ted and Laurie's awareness of any and every change in Pascal's behavior makes it highly unlikely they'd miss any illness, their supersensitivity and lack of confidence lead them to assign extreme meaning to even the most insignificant behaviors. In their enthusiasm to respond to everything he does, they wind up overresponding. Like most owners they expect to indulge a young pup, but are horrified to discover that no amount of round-the-clock pampering seems to pacify a tyrannical dog.

Let's follow the development of one particularly troublesome behavior, the aforementioned vomiting, and see how such frustrating connections between emotion and body language can occur. First, let's examine the behavioral significance of the display. Because dogs gulp their food, Mother Nature and evolution have kindly provided them with the ability to vomit quite easily. Obviously, if a dog doesn't bother tasting what it eats, it behooves it to be able quickly to expel anything disagreeable or inappropriate. Furthermore, wild dogs feed their offspring via the process of regurgitation: Adults consume prey at the site of the kill, then return to the den and vomit it up for the pups. Transporting food in this manner is not only more energy-efficient than dragging a carcass back to the den, it keeps the pups' food from being stolen by larger scavengers. Because the regurgitation response became linked to pup survival, the behavior was perpetuated. In addition, we already discussed how more fearful dogs are prone to hypermotility, a phenomenon that further sensitizes the already delicately triggered vomiting mechanism.

Even though Gloria's past and the Crabtrees' current evaluations might convince us that Pascal is nothing more than a dominant, despotic little beast, the opposite is actually true. The young pup's shyness causes it to gaze imploringly at his owners when they place his food dish on the floor, and his timidity makes him squeal in surprise—not pain—when Ted accidentally bumps him. Obviously, when his owners change the food, cuddle him, or otherwise respond positively to his "sadness," they wind up rewarding the very behaviors they want to extinguish.

From such humble beginnings do canine monsters evolve, until the concerned owners are spending most of their waking hours trying to make a fearful dog happy. Think about that: Suppose you're terrified of thunderstorms but adore hot-fudge sundaes. Regardless of how many hot-fudge sundaes you consume during a storm, does your sated appetite eliminate your fear of thunder? No; it may distract, but once you've eaten your fill, the storm will still be rattling the windowpanes. Furthermore, the greater your fears, the less hungry you'll be and the more difficult to distract.

So as the Crabtrees prepare to leave for the evening, the unusual activity makes Pascal uneasy. He observes his owners anxiously. Unable to stand the "sad" look, they pick him up and cuddle him, all the while murmuring an endless stream of conciliatory chatter: "Poor, poor Pascal, don't be so sad. We won't be gone long." Just before leaving, Laurie hugs the terrier tightly for the umpteenth time. In addition to the hard squeeze and her almost frantic reassurances, he also gets a big whiff of her strong perfume.

As she puts him down, he gags, a signal his already hypermotile system interprets as an invitation to chuck up dinner—and all those treats the Crabtrees lavished on their sad dog to make him happy and assuage their guilt. In this case, the diagnosis that "Pascal has a nervous stomach" only makes matters worse. Now the Crabtrees' definition of the normal Pascal expands to include a preference for unhappiness and a tendency to vomit when upset. Whenever the dog vomits, they feel they've failed yet again in their efforts to make him happy.

Clearly the goals and definitions in such a relationship need a major overhaul. Like Gloria, the Crabtrees need more workable goals and a new understanding of normal to replace the ones forcing them to shoulder a burden of ownership rather than the pleasure of dominion. If they can settle for wanting a good relationship with their dog rather than wanting to be perfect owners, Ted and Laurie will be able to recognize the contributions both owner and pet make to every positive and negative interaction. By spending a few weeks simply observing Pascal, instead of constantly waiting on or reacting to him, they quickly discover that it's not normal for Pascal to act exuberantly happy all the time. Left alone, he's a shy, quiet little dog who prefers to remain close to his owners but doesn't necessarily appreciate being fussed over. At first Laurie is miffed to discover he sleeps just as soundly if not more so without her elaborate nightly cuddling ritual; and Ted experiences a twinge of irritation when he realizes that Pascal actually grows less fearful if his behavior is ignored than he ever did when he received lavish consolation.

Since the Crabtrees got swept up in a host of problems because they lacked confidence, they find it hard to accept the simple explanation that their overindulgence causes more problems than it cures. Such an explanation only decreases their confidence further. Doesn't it make more sense for them to create changes that will increase their confidence as well as that of their timid pet?

Once the Crabtrees accept the fact that most of Pascal's seemingly sad displays arise either from his normally quiet nature or from his fear, they initiate daily training to eliminate the latter. Because their desire to be perfect owners helped them concoct an imaginary ever-jubilant canine ideal that exceeds normal canine personality and certainly lies beyond the shy Pascal, redefining their relationship in ways that include rather than exclude Pascal's norm produces immediate benefits. By abandoning their previous goal as both unobtainable and abnormal and embracing the quieter and more balanced relationship as a more beneficial norm, Laurie and Ted can relax. As soon as they relax and stop worrying so much about whether Pascal is happy or sad, the dog's defects begin to disappear, and the whole relationship quickly stabilizes.

While the relationship doesn't contain those exhilarating highs the Crabtrees originally believed critical to ideal canine interaction, neither does it plunge them into those devastating lows during which they must scramble to alleviate Pascal's sadness. Furthermore, as Pascal responds to his training and his confidence grows, a different kind of high emerges. In contrast to the sudden bursts of enthusiasm Ted and Laurie previously recognized as evidence of a sound relationship, they now become aware of a continuous mellow glow arising from their own and their pet's new confidence and contentment in each other's presence.

PSEUDO-SORROW, PSEUDO-GRIEF

No discussion of sadness, sorrow, and depression would be complete without exploring death. As we have seen, linking certain behaviors to sadness can easily lead to guilt, overindulgence, and an unstable human/canine bond. "But," cries the dog-loving owner, "I know my dog experiences true sorrow!" Again, while the behaviorists would disagree, such owners will point to Greyfriar's Bobby and all those dogs that mope around when their masters leave them temporarily or permanently.

A quick objective analysis of this phenomenon based on what we've already discussed tells us that a social animal will react to even small changes in pack structure. The smaller the pack, the greater the void created by the disappearance of a more dominant member and the greater the sense of vulnerability on the part of those who depended on that individual. Dogs who pine away when separated from their owners are generally highly dependent animals who formed close attachments to only a few people or even to just one person. While we noted that these intense relationships may provide the owners with ample opportunities to while away the day ministering to the dogs, when such an owner leaves for more than the shortest time, the dog suffers psychologically, often stops eating and drinking, and falls prey to any one of a number of unhealthy states awaiting the low-resistance individual.

Ironically, dependent dogs often evolve because of their owners' acute awareness that their pets probably won't outlive them. As our dogs age, many of us consciously or subconsciously begin grappling with the reality that our beloved pets are going to die. Like many owners, Nancy McKuen dreads the day her Irish water spaniel, Shamrock, will die. Even when he was a young pup, Nancy underscored every positive interaction by noting, "He's so wonderful, I simply can't imagine life without him." Understandably, as Shamrock ages, Nancy's fears increase proportionately—until she spends a tremendous amount of time trying to make conditions perfect, ostensibly so that he'll be healthy and happy "as long as possible," but in reality in the hope of making him immortal. She feeds him only the best of food (a specially blended, expensive diet she must drive fifty miles to purchase) and supplements this with an elaborate mixture of vitamins, minerals, and herbs. She grooms, bathes, walks, runs, and plays with her dog according to a complex schedule she adapts from a book on human longevity. Her interactions with Shamrock become so intricate and time-consuming that he rarely comes into contact with other people or animals.

One day it happens: Fate separates Nancy and Shamrock. No, the dog doesn't die of old age, but simply becomes frightened and bolts into the woods when Nancy lets him out of the car for some exercise at a secluded rest stop. Or maybe she must kennel him because a family crisis necessitates an immediate cross-country trip and she can't afford to take him. Or perhaps Nancy becomes ill and must be hospitalized. Or perhaps she even dies. Can a dependent pet like Shamrock make a speedy, stable transition to a new environment and owner? How willing would you be to adopt an adult dog with all Shamrock's complex requirements? Like many dependent owners who tie their pets to them by reinforcing the animal's needs or even creating new ones, Nancy confronts the tragic paradox of this approach. To be sure, such owners may gain satisfaction from their pets' devotion and dependence, but they must also accept that if anything happens to them, that beloved pet will probably be doomed to a long and often agonizing period of readjustment, one over which others may have neither the time nor the patience to preside. Whatever pride and security these owners glean from being able to say, "Shamrock and I are so close, I know he'd die if anything happened to me," they must balance that with the awareness that such a romantic, self-serving observation may one day come true. The dependence so heavily reinforced and magnified in the name of love and so contingent on the owner's presence may indeed sound the death knell for the dog—either through illness or through euthanasia—when the two are permanently separated. When such animals pine away and die, only the unknowledgeable human can romanticize the event and attribute it to grief and sorrow, and the owner's creation of such a relationship to his or her great love.

At the opposite end of the sorrow spectrum, independent owners perpetuate another phenomenon that can spell trouble for the dog whenever owner and pet are separated for any extended period. Because these owners reinforce the differences between human and canine as something to be mastered and subdued, few feel compelled to step in and care for or assume ownership of these pets if separation occurs. Whereas the highly dependent pets are one-man dogs by virtue of the great amount of care they often demand, the independent owner creates a one-man dog by virtue of all the (real or imaginary) need for human control they believe the dog requires. Only other independent owners who view such a dog as a challenge can accept or want such a relationship. Because the independent owner uses his or her orientation literally to *exclude* the dog from other relationships, any new owner must first "conquer" the dog, win its trust, then establish new rules once again designed to exclude others. To observe this scenario, we need only peruse the majority of books and movies dramatizing human interactions with wolves.

While the new owner may revel in the fact that he or she has succeeded in "retraining" the beast with great skill and patience, we can't overlook two troublesome facts:

- The animal has to undergo physical and/or psychological "conquest" and "retraining."
- The dog may not enjoy a "better" life; if something happens to the new owner it will probably be "whipped into shape" by yet another with an independent orientation. Barring the presence of such an individual it will be euthanized because it's untrainable.

So once again, insecure owners who can't bear the thought of living without their dogs (or of their dogs living with anyone else) create pets whose chances of surviving any kind of life without them dwindles to almost nil. Anyone who would point to the animal's sorrowful condition when separated and call that a sign of the owner's great love projects values that find no place in a bonded relationship.

TRUE SORROW, TRUE GRIEF

Much of the joy inherent in a bonded relationship comes from the relationship's delightful surprises. We noted how our most courageous protectors are our most happy-go-lucky pets and how well-bonded dogs seem to learn so much more easily because their owners believe they want to learn. While such wonderfully positive emotions do spring from these facts, they also offer owners quite practical and comforting reassurances during times of separation.

Taking a slightly different view of sorrow and grief in our relationships also offers some pleasant surprises. Recent scientific studies indicate that tears of sorrow contain unique chemicals that are the by-products of sadness. People who are able to "cry it out" do get over their sadness sooner because they eliminate these by-products and their negative effects more quickly from their bodies. Considering the dog's incredible sense of smell, this could certainly explain why well-bonded pets remain sympathetically (empathetically?) near when we're shedding unhappy tears, but are totally unmoved by those we emit when we peel onions. Although we may hear tales of dogs whose watering eyes seemed to indicate sorrow, we can find no scientific evidence to support such occurrences.

If we view sorrow and grief as cleansing, transitional emotions that allow us and our dogs to deal with traumatic change, including separation, we can use them to our benefit. Think of sorrow and grief as a means of quickly traveling from one unstable situation through a disorienting middle ground to a new stability. Obviously the shorter the period of intermittent disorientation, the more quickly we stabilize.

This brings us to the second surprise that attends the bonded view of sadness and separation: By choosing to relate to our pets in such a way that they could be happy with anyone, we simultaneously create a bond that holds us close together *and* guarantees that we can both survive being apart.

If for some reason bonded dog and owner separate, the dog fairly quickly gravitates toward a new owner. To be sure, it may experience an initial period of disorientation—grieving, if you will—as it accommodates to a new environment and/or pack structure. For owners who haven't provided for their pets in the event of their absence, this may even include a period of isolation in a boarding kennel or shelter until a new home can be found. The more bonded and stable the pet, the more quickly and happily it will make any necessary adjustments. Anyone who's ever worked in a shelter can instantly recognize these more adoptable bonded animals. Even in areas overpopulated by dogs, people appear almost magically to adopt such animals or campaign on their behalf. Some end up as cherished shelter mascots or official greeters. In short, they almost invariably find good homes.

What about those stories in which separated pets endure all sorts of horrible experiences? So often we forget that these events fascinate us because they're so *rare*. Furthermore, as we read of the dog's agonies, we lose sight of the fact that a dog that began with a stable relationship almost invariably achieves one in the end. The lost dog is reunited with its loving family, the old hermit's beloved hound is befriended by the troubled teenager. Like, once again, begets like.

If our dogs attest to the strength of their love for, and their bond with, us by the ease with which they accept others, what does that tell us about our feelings when we separate from our dogs? Certainly we, too, will suffer an immediate sense of disorientation and loss. Some owners may immediately rush out and draw a dog to them, while others respond oppositely, repelling even the idea of another pet: "No other dog can ever replace Chris."

Whether the bonded owner gets another dog or not, the sense of loss and sorrow gradually gives way to fond memories. Although such an owner may never own another dog, he or she never cites the quality of the relationship with the first dog as the reason. Contrast this to the owner who clings to every detail of the events preceding and following the demise of the relationship with the dog. "I loved him too much, I couldn't bear to go through that again," sobs Nancy McKuen a year after she has Shamrock euthanized with inoperable bone cancer at age fourteen. This is understandable if all Nancy chooses to acknowledge from her fourteen years with Shamrock are the *three weeks* of mental anguish that surrounded the diagnosis of his condition and her choice to put him down. Having given that memory such a powerful negative charge, she would indeed be a fool to put herself through such pain again.

Each person reacts to the death of a loved one in his or her unique way. Two friends of mine displayed common but quite different approaches to euthanasia. The husband and wife had separated after seventeen years of marriage, during which they had raised a handsome Irish setter, which remained with the wife. When the dog's health finally failed, the wife couldn't bear to make the inevitable decision, despite the great deal of discomfort the dog was experiencing. The husband called for advice. "What should I do?" he asked. "Should I just wait for the dog to die? Should I insist on taking him to the vet's myself, regardless of what Shirley says?" Because his wife had lost her parents a year earlier and was still too wrapped up in her grief to handle yet another death of a loved one, I suggested he offer to take the responsibility on himself. He did, and she accepted, although it was quite difficult for her. Later he told me about those final moments he shared with his pet:

"I'd never had a close friend or relative die, so when the vet asked if I wanted to hold the dog while she put him to sleep, I was afraid I couldn't handle that. But I'd seen him come into the world and figured I ought to see him out of it, too. I'm glad I did. When he fell asleep, then grew cold in my arms, I cried like a baby, but I left the vet's office feeling good about myself and my life with that dog."

In this case the husband found great solace being with his pet during those final moments, whereas his wife realized that being there would only exaggerate her own feelings of grief. At such a time, above all others, each owner must act in accordance with his or her own wishes and needs; othewise the acceptance of the loss will be marred and protracted by the presence of guilt.

The way in which we and our dogs respond to loss should be the final tribute to the quality of our relationships. I'll never forget one such tribute. A dignified old farmer, tears streaming down his face, stood staring at his dog's lifeless form on the table between us. "Well, there goes the last of Mickey's bad habits," he said with a faint smile, undoubtedly referring to his beloved Mickey's singular vice: sneaking under the covers after his master had fallen sound asleep. Was my client taking what little pleasure he could from this wretched occasion by noting that he no longer would have to endure this misbehavior? Not at all. What he simply and beautifully communicated was that in death Mickey became perfect in every way because his owner chose to remember all the experiences they shared as good.

The value of sorrow and grief doesn't depend on either their intensity or their duration. For the bonded owner these emotions serve as a vehicle for traversing the rocky ground between the end of one era and the beginning of another, which may or may not include another pet. To evaluate our relationship with a previous pet or prejudge any future one based on perpetuated or exaggerated memories of the brief period of loss denigrates the former relationship and dooms any to follow. Sorrow and grief should never be terminal emotions, but rather physiological and psychological cleaning agents, a means of sorting out old memories and hanging on only to the best. The more solid the relationship, the less sorting we need to do, the quicker we accomplish the transition. As my client noted so well, it doesn't take much to make an almost perfect dog a perfect memory.

12

THE CHOICE

IN THESE PAGES we've explored a wide variety of body-language displays and their concurrent and elicited human and canine emotions. Regardless of whether we spoke about something so unemotionally cut-and-dried as the behaviorists' description of dominant canine displays or so intimate as love, we were never far from the realization that accepting or rejecting any of this information was and is a matter of choice. The Sullivans can maintain their slavish anthropomorphic orientation toward Bit O'Honey, Richard Wilcox can continue rearranging his life around Muffin, Cathy McCaffrey can continue jumping when Czar's alarm barks tell her to jump—but it doesn't *have* to be that way. These owners don't have to relate to their dogs as fur-coated babies, dependent wimps, or vicious protectors. Nor do more independent owners like Lou Rutherford and Roy Ringhausen *have* to respond to their pets as though they were some alien, recalcitrant species that must be whipped into shape, any more than the Terry and Pamela Pedersons of the world need be part of their dog's pack rather than vice versa. All the displays we've studied represent choices owners make, sometimes unknowingly, sometimes with full awareness of the effect their interpretations have on their dog's behavior and what their own body-language displays communicate to their pets.

In addition, we took a look at the objective data offered by the scientists who prefer to strip all emotion from their studies of canine behavior. This is a choice they make, most likely for a very practical reason: Emotions are so numerous and subjective that their incorporation into any system of evaluation adds an unwieldy number of variables to the process. However, just because the scientific community has chosen to exclude emotions from their data doesn't mean that emotions don't exist. All it means is that the presence or absence of emotion isn't acknowledged *in their data.*

While such reseach provides us with highly objective and valuable insights into our dogs' behavior, it also requires that we make another choice: Should we accept this view as indicative of our dog's total range of expression, or should we add the "unproven" dimensions of canine thought and emotion? For most of us the idea that our pets aren't thinking, emotional beings is so incomprehensible that we automatically incorporate these parameters into the relationship without giving them a second thought. However, by recognizing that their inclusion *is* a choice, we retain the freedom to disregard those thoughts and emotions we attribute to our pets that erode our relationships. Whereas the scientific approach says, "Get rid of all emotions," this bonded approach says, "Get rid of the ones that don't work for you and your dog." When Loretta Letesla views Chiquita as a wolf in Chihuahua's clothing, when the Mathesons believe Fezziwig bored to tears, when the Carlisles perceive jealousy in Dickins's unemotional pack behavior, the owners' choice to project these human emotions on their pets serves no positive purpose. Not only don't such interpretations lead to any solutions, they compound the problem by adding negative emotions to the already-present negative body-language display.

For those who prefer more predictability and objectivity in life, the idea of accepting or rejecting certain emotions at will and according to circumstances may be unnerving. If you feel that way, by all means don't force yourself to recognize emotions where you believe none exist. On the other hand, beware the trap that insidiously ensnares many "unemotional" owners. They find people who speak of the reciprocated love, devotion, happiness, or sadness with their canine companions foolish and immature, but don't hesitate to become angry with their own pets and accuse them of being deliberately spiteful and mean. In such a way they divorce themselves from all the positive emotions available in pet ownership and accept only the most negative. If you don't believe dogs experience positive emotions such as love or happiness, then do yourself and your dog a favor and don't attribute *any* emotions to it. While some may argue that the result will be a rather insipid experience for both human and canine, at least it will be consistently so and will spare the dog the myriad problems that attend inconsistently evoked negative emotions.

Doesn't similar inconsistency result when we choose only to acknowledge positive emotions in the relationship? Theoretically this is true. However, in both cases the owners offer their pets relationships governed by one of two emotional states. Those who consider positive emotional interaction frivolous offer their pets only negative or neutral emotional experiences. Contrast this to the owner who reacts neutrally to the former's negative events and shares positive interactions the rest of the time. To be sure, it's a highly subjective gamble each of us accepts or rejects and reminds me of a statement made by a favorite philosophy professor regarding that unproven state called Heaven: "If we can't be sure whether a heaven exists, we've nothing to lose by pretending it does and striving to achieve our full potentials and improving the lot of our fellowman. If it doesn't, we still succeed in living a full and beneficial life. If it does, we're assured even greater glories in the afterlife." If you're not sure whether your pet is capable of experiencing and expressing emotions, why not choose to transmit and receive only the best? If our dogs do lack this capacity, we succeed only in making ourselves feel better. If the capacity exists, even in a different form, we expand the scope of our interactions tremendously.

Above all, bear in mind that the manifested form of choice is change. Each choice we make regarding our pets opens new paths of experience, which offer more opportunities to learn and grow. In less than a year's time, Ruth Lively went from viewing Nova as the perfect stay-at-home companion to seeing him as a grieving pet punishing her for leaving him alone, to accepting him as a more submissive animal manifesting isolation stress via minor misbehaviors. Imagine how her relationship would have suffered if she had been unwilling or afraid to make the choices that precipitated these changes because she believed each was an absolute that could never be changed again. As it was, she took each new experience with her pet, evaluated it, and changed or didn't change it, according to her needs at that time. While some may feel that she put herself and Nova through needless emotional trauma, her confidence in her relationship with the terrier allows her to look back on all their times together with no regrets.

Similarly, when Lou Rutherford summons the memory of Merlin and the chewed catcher's mitt, he doesn't recall his anger, guilt, and violence. What he remembers most vividly is the power of the emotion that "dumb dog" was and is capable of eliciting in him. That was his first hint of the magnitude of the potential bond between him and Merlin, and now that it manifests positively, he finds it difficult to remember it was ever any other way. But there was a time when Lou wanted to play caveman to Merlin's wolf, when he wanted his dog to submit and feel guilt for all his transgressions. However, Lou changed and changed his dog time and time again until he achieved a relationship that made him feel good about himself and his pet.

Nor is Lou unique; each one of us harbors pockets of dependent and independent beliefs, ready to burst open when our dogs misbehave or otherwise complicate our lives and we'd prefer to see ourselves as victimized by, rather than responsible for, our pets. But each time we recognize the presence of these beliefs and our choice to accept, reject, or change them, their ability to dominate our relationship declines. As our domination by negative, nonproductive beliefs declines, our ability and willingness to accept dominion over our dogs increases proportionately.

Interpreting our dogs' body-language displays, assigning

them emotional motivations, and then responding to them, offers us infinite opportunities to temper and rework our relationships with our dogs and forge a solid yet flexible bond. The ultimate body-language display is our willingness to change, and change our dogs if necessary, to create such a bond; the most perfect emotion is the joy we feel and the love we express when we make a conscious choice to do so.

INDEX

A

acceptance
  of canine time, 204, 216
  guilt and, 101–102
  isolation behavior and, 186–
    188
  patience and, 53
  of problems, 64, 73, 77–78
  of spite and hate, 235–236
aggression
  balanced with fear, 157
  as behavioral state, 47
  factors influencing, 160–163
  fear and, 152–173
  hostility and, 48
  self-evaluation of, 46–48
alienation response, 95–96
anger, self-evaluation of, 46–48
anthropomorphism
  dependent species orientation,
    28–32
  "inhuman" canine behaviors
    and, 30
  problem solving and, 74–75

attention response, 94–95
awe of dog, 52

B

balance
  of dominance and submission,
    170
  of fear and aggression, 157
ball-dependent displays, 40
behavioral problems
  recognizing, 63–64
  solving, 62–84
behavioral states
  defined, 41
  frustration, 44
behaviorist's view
  of consistency, 24
  defined, 19
  of depression, 252–253
  of devotion, 52
  of fear and aggression, 49,
    154–158